Foundations of Engineering Mechanics

Babitsky · Shipilov

Springer

Berlin
Heidelberg
New York
Hong Kong
London
Milan
Paris
Tokyo

V.I. Babitsky · A.V. Shipilov

Resonant Robotic Systems

Translated by N. Birkett

With 110 Figures

Springer

Series Editors:
V. I. Babitsky
Department of Mechanical Engineering
Loughborough University
Loughborough LE11 3TU, Leicestershire
United Kingdom

J. Wittenburg
Institut für Technische Mechanik
Universität Karlsruhe (TH)
Kaiserstraße 12
76128 Karlsruhe
Germany

Authors:
V.I. Babitsky
Department of Mechanical Engineering
Loughborough University
LE11 3TU Loughborough, Leicestershire
United Kingdom

A.V. Shipilov
Pricewaterhouse Coopers LLP
Global Risk Management Solutions
PwC Suite 3000, Box 82
Royal Trust Tower TD Centre
Totonto, Ontario M5K 1G8
Canada

Translator:
Nicholas Birkett
28 Grangefields Drive
LE7 7NB Rothley, Leicester
United Kingdom

ISBN 978-3-642-05563-8 **e-ISBN 978-3-540-36380-4**

Cataloging-in-Publication Data applied for
Bibliographic information published by Die Deutsche Bibliothek.
Die Deutsche Bibliothek lists this publication in the Deutsche Nationalbibliografie;
detailed bibliographic data is available in the Internet at <http://dnb.ddb.de>.

Springer-Verlag Berlin Heidelberg New York
a member of BertelsmannSpringer Science+Business Media GmbH

http://www.springer.de

Printed in Germany

Cover-design: de'blik, Berlin
Printed on acid-free paper 62 / 3020 hu - 5 4 3 2 1 0

Preface

In 1984 in the Institute of Machine Studies of the Russian Academy of Sciences a prototype of the new manipulative system MARS with record-breaking characteristics was completed and demonstrated. The machine weighed 80N and was able to manipulate accurately in a cylindrical co-ordinate system with parts weighing up to 5N, moving them quicker than conventional industrial counterparts which usually weighed four times more. For all of that its miniature electric motors and control system consumed less than 20 W of power.

Manipulation auto-resonant system (MARS) used for its movement the natural modes of oscillation of electromechanical modules specially designed as self-sustaining oscillating systems. The oscillation of an actuator in each module was interrupted with the help of electrically activated latches at the points of prescribed fixation when the speed of the module actuator was close to zero. This was performed by signals from a control program.

The inertia forces of the moving parts of the manipulator were equalised at every instant by springs installed in every module, and the electric motors replenished only the energy lost due to friction or useful work of the oscillatory systems. The adaptive control system was able to identify the unknown mass of the payload and then regulated the excitation of each module displacement according to the received estimate in order to bring it to the fixation position with a minimal distortion of the natural movements.

The MARS resonant robot system was widely acclaimed in Russian technical literature and mass media. Under the influence of this success a series of new devices using similar principles was proposed and developed. This was followed by research of autoresonant mechatronic systems, which were used in these devices. Due to this some new nonlinear problem were formulated and solved.

The subsequent complication of the Russian political and economic situation interrupted this technological development. However, with an increase in the capacity of intelligent control systems there are new efforts being made to bring the above principles into technological applications. This has stimulated the authors to sum up the results and experience on research and development of resonant robotic systems in this book.

For the convenience of readers we present in the book the foundations for the dynamics and control of resonant robotic systems which will help both students and professionals to understand the mechanics of these systems as well as the principles of their analysis and design. The latter is supplemented with design schematics for some new inventions. Not all of them have been technically used and tested but we hope that acquaintance with these ideas will stimulate a new generation of engineers to develop more perfect and sophisticated autoresonant mechatronic systems using the latest technological achievements.

The authors wish to acknowledge our colleagues in the Institute: T. Akinfiev, M. Chitayev, A. Kovaleva and N. Makarov as well as the managers and designers of Moscow enterprises: V. Belov, N. Kriukov, V. Yurchenkov, and S. Pavlov whose efforts and support made it possible to accomplish the technological development of resonant robotic systems. We are grateful to the translator of the manuscript Mr N Birkett and the staff of Springer for the production of this book.

V. Babitsky, A. Shipilov
Loughborough, Toronto

Contents

Chapter 1. Application of resonant systems

§1.1 Load relief and balancing of robotic systems

Before solving problems of synthesis of an industrial robot as an integral dynamic system, the creation of separate drives that ensure partial motions of the robot must first be dealt with. In this case, we shall assume *partial* motion to mean programmed motion of the robot in which the operation of one of the drives[1] causes only one of the generalised coordinates of the robot to change [67]. Dynamic synthesis of an industrial robot is based on obtaining mutual co-ordination of the drives taking account of their dynamic interactions as the movements occur simultaneously with several degrees of freedom.

Let us look at the forces in the transmissions for the separate degrees of freedom of an industrial robot. The *transmission* includes all the elements of the transfer mechanism that connect the actuator to the driven member. For a transmission mechanism we can write

$$\begin{gathered} F_r + F_w + F_b + F_i = 0, \\ D + M_r + M_w + M_b + M_i = 0 \end{gathered} \tag{1.1.1}$$

Here F and M are forces and principal moments of force; r represents resistance; w represents the weight of components and load; b represents the reaction of the base; i represents the inertia of components and load; D represents the driving torque.

One of the fundamental methods for improving the dynamic properties of robots is to reduce the load level in the drives. In practice, methods for load relieving and balancing mechanisms are used [67, 93, 107].

The aim of balancing a mechanism is to remove any variable effects caused by vibration of the complete mechanism from the base itself and from the housing elements to which the base is attached. A *balanced mechanism* is one in which the resultant vector of forces and the principal moment remain constant during assigned motion of the mechanism components. The conditions for complete balance of a mechanism take the form

$$\begin{gathered} F_r + F_w + F_i = const \\ D + M_r + M_w + M_i = const \end{gathered} \tag{1.1.2}$$

Obtaining exact satisfaction of the conditions (1.1.2) is a complex theoretical and engineering problem. Thus, in practice, for high-speed mechanisms it is only

---

[1] We use *drive* as the generic term for a dynamic system, which is composed of an actuator (e.g. electric motor, hydraulic or pneumatic driver), a reduction system and a transmission system.

possible to obtain partial balance. The body is protected only against the effects of inertial forces and their moments caused by moving components.

A mechanism for which the principal vector of the inertia forces

$$F_i = 0 \tag{1.1.3}$$

is called a *statically balanced mechanism.* As such a mechanism operates, the centre of masses of all its moving components remains stationary.

If the principal moment of the inertia force

$$M_i = 0 \tag{1.1.4}$$

then the mechanism will have *moment balance.*

If both conditions (1.1.3) and (1.1.4) are satisfied simultaneously then the mechanism is a *dynamically balanced mechanism.*

There are many methods for balancing mechanisms. The most straightforward of these is the *substitute masses method* [11, 93]. For example, in a plane mechanism a system of localised masses $m_1,\ldots,m_n$, which has the same mass m, the same centre of masses and the same moment of inertia as a substituted link, is called a substitute masses system.

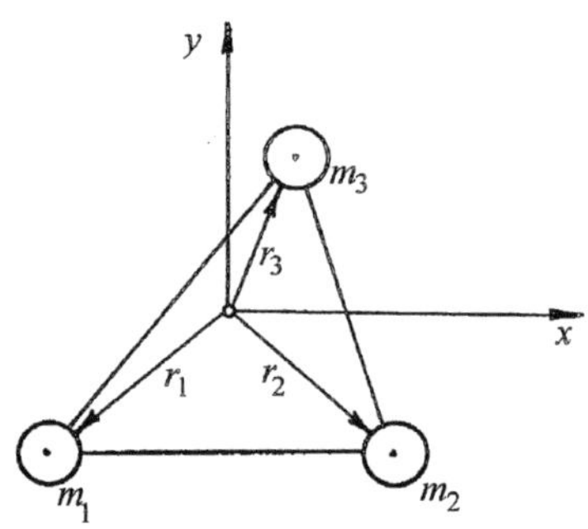

Fig. 1.1.1

Setting the origin 0 of the xy coordinate axes at the centre of mass of a link (Fig. 1.1.1), we can then write for the three substitute masses

$$\begin{aligned} m_1 + m_2 + m_3 &= m, \\ m_1x_1 + m_2x_2 + m_3x_3 &= 0, \\ m_1y_1 + m_2y_2 + m_3y_3 &= 0, \end{aligned} \tag{1.1.5}$$

$$m_1(x_1^2 + y_1^2) + m_2(x_2^2 + y_2^2) + m_3(x_3^2 + y_3^2) = J_0 \tag{1.1.6}$$

When conditions (1.1.5) are satisfied the mass disposition is called *static.* In this case the principal vector of inertial forces of the link is equal to the principal vector of inertial forces of the substitute system.

The substitute mass disposition is known as *dynamic* if conditions (1.1.5) and (1.1.6) are satisfied simultaneously. In this case the principal moments of inertial forces are equal.

If the abscissa is parallel to the line connecting the centres of masses m_1 and m_2 so that $y_1 = y_2$ then the substitute masses will be defined by the equations

$$m_1 = m\frac{x_3y_1 - x_2y_3}{(x_2 - x_1)(y_1 - y_3)},$$

$$m_2 = m\frac{x_1y_3 - x_3y_1}{(x_2 - x_1)(y_1 - y_3)},$$

$$m_3 = m\frac{y_1}{y_1 - y_3}.$$

When counterbalances are installed the construction of the transmission becomes more complex. The natural frequencies of the system are reduced and due to an increase in the variable component of the equivalent moment of inertia the vibration of the drive is increased [107]. In reality only approximate balancing can be applied, by balancing the first harmonics of the principal vector and the principal moment of the unbalanced forces [93].

Dynamic balancing has not been used widely with robotic systems because flexible programs operate manipulators with predominantly transient motion conditions. One of the main drive tasks is to supply exacting start-up and braking conditions to the operating components. Installation of counterbalances increases the equivalent moment of inertia of the moving components, reduces the operating speed and increases the drive load.

Improvements in robot dynamics are more fully achieved by *systems for load relieving the drive.* These consist of methods to reduce dynamic and static loads in the separate elements of the transmission and in the complete drive. The load relief devices ensure that those parts of the transmission subjected to significant pulsating loads are isolated. As a result the friction forces in kinematic pairs are reduced and the endurance, reliability and drive efficiency are all increased. This creates a basis on which material usage and energy consumption can both be reduced and the operating speed of the robot can be increased.

It must be emphasised that the problem of load relief, i.e. compensation of the loads on the drive components, is not the same as the problem of balancing the reactions that occur on the drive support. Load relieving systems to provide a decrease in the maximum value of the reactions in kinematic pairs cause not only spatial redistribution of the internal forces acting on the drive but also redistribute them in the motion cycle. As a result the cycle time can be reduced. The drive remains unbalanced on its support, and the load relief system can become a significant source of vibration for the drive.

Active, passive and combined systems of load relief of drives are all used in industrial robots. *Active load relief systems* produce direct load compensation by

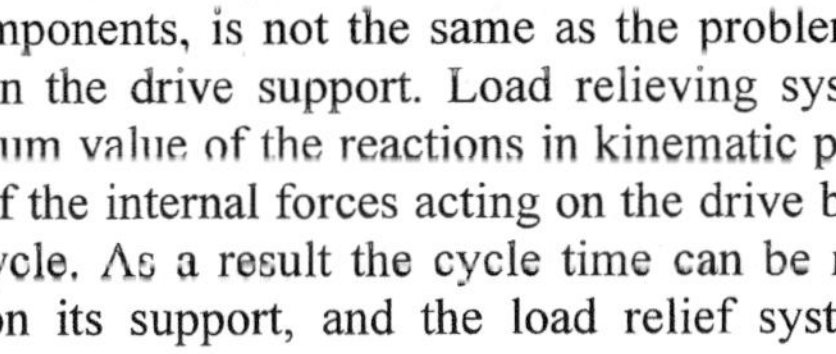

Fig. 1.1.2

using information about the phase coordinates of the operating element. An electro-hydraulic drive is mainly used in load relief devices because it can relieve significant forces without needing an additional transmission mechanism.

Fig. 1.1.2 shows schematically the active load relief system for a robot drive. Apart from the basic driver D_1 connected to the operating component OC by means of the transmission TR, the system contains a supplementary load relief driver D_2 that is set to operate at the instant when the transmission reaches its maximum loading. To match the control voltages u_1 and u_2 of drivers D_1 and D_2 the control system CS is operated according to the phase coordinates φ and $\dot{\varphi}$ of the operating component using the feedback system FS.

Passive load relief for the drive is provided by using counter-balances or mechanical energy accumulators, which may store either potential or kinetic energy. Fig 1.1.3 shows the energy store *ES* acting on the operating component

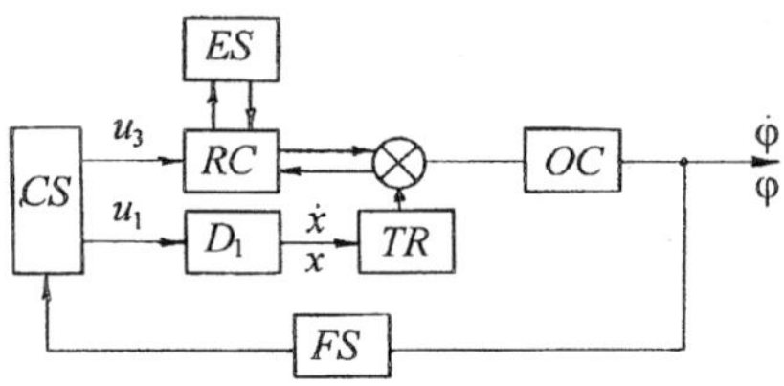

Fig. 1.1.3

OC of the robot in parallel with the basic drive. The regulating component *RC* controls the process of charging and discharging the accumulator *ES*. Passive load relief makes it possible to decrease not only the required power of the drive, but also energy consumption of the whole robot. There are also disadvantages since the total mass is increased and the structure of the robot mechanical system becomes more complex.

High efficiency industrial robots can be obtained by using *combined load relief systems* (Fig 1.1.4). As distinct from traditional control systems using feedback, in

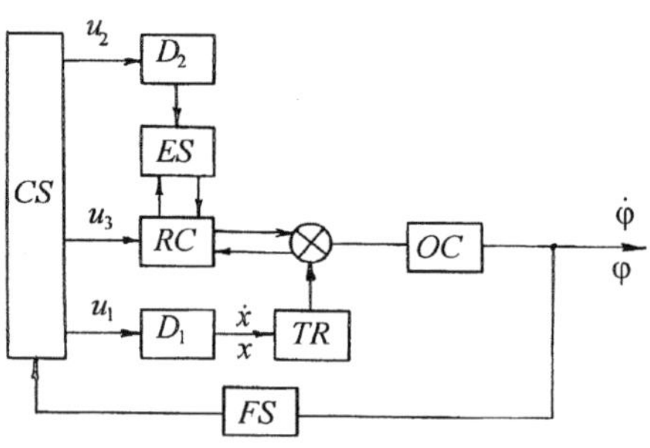

Fig. 1.1.4

this system the basic driver D_1 implements programmed motion $\varphi(t)$ and the feedback control effect is applied to the transmission through the energy store *ES* that is fed by the supplementary driver D_2. The energy store in a combined load relief system is usually a pneumatic system. Depending on the operating program and the phase coordinates φ and $\dot{\varphi}$ of the operating component, the driver D_2 adjusts the pressure required for the pneumatic accumulator output with the help of regulating component *RC*. Combined load relief reduces the power requirement of the primary drive and the total power consumption of the robot; it also provides somewhat better load relief for the drive.

During operation both static and dynamic forces are experienced by the robot drives. In different forms of programmed motion, static or dynamic loads become more dominant. Since relieving the drive of all the loads simultaneously presents a severe technical problem, load relief systems that can cope either with static or dynamic forces are usually built in practice.

The condition of *static load relief for a robot* from the effects of forces due to component weight and transferred load is the independence of the potential energy of the robot from its positioning or generalised coordinates. Static load relief of a robot is obtained by a rational masses distribution or by using elastic components

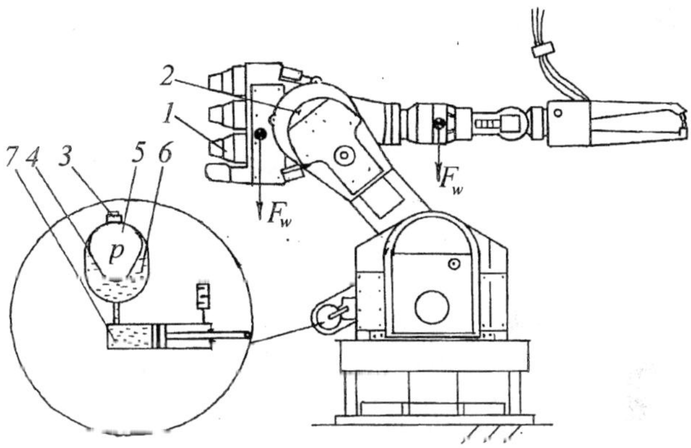

Fig. 1.1.5

that store potential energy. In the case of rotational partial motion the masses are distributed so that the centre of masses of the rotating system is a maximum close to the axis of rotation. If all the kinematic pairs of the manipulator are rotational, static load relief can be achieved by using static balancing components. Supplementary counterweights are used for this, as in the robot type *IRb-6 (ASEA)*, or by arranging the manipulator elements suitably, as the electric motors *1* in the robots *KUKA* (Fig. 1.1.5) and *SMART*.

Static load relief of drives undergoing partial sliding motion can be produced using counterweights flexibly connected to the driven element (Fig. 1.1.6). Note that complete static load relief of a robot by a suitable arrangement of the masses is not possible in most cases because of the changing weight of the transferred loads. Moreover, the presence of supplementary counterweights and elastic links

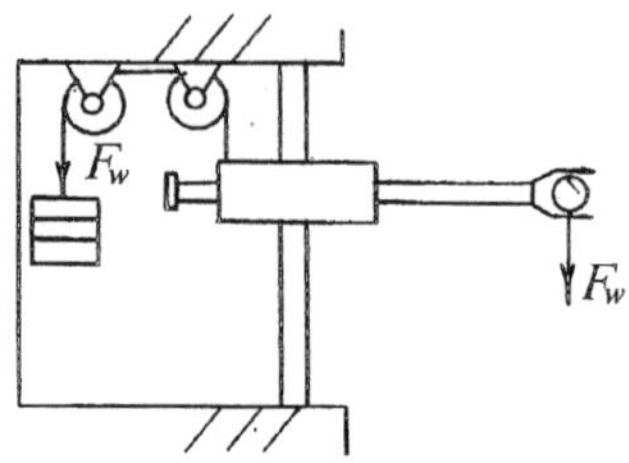

Fig. 1.1.6

cause the dynamic performance of a robot to deteriorate.

Elastic elements are the most widely used components to store energy in the load relief systems of drives, which are acted on by the weight of moving components. For this purpose, either springs or compressed gases can be used. Pneumatic-hydraulic accumulators, in which the inert gas pressure reaches $1.8 \times 10^7 Pa$, possess the highest specific energy output. The efficiency of using a spring load relief or a pneumatic load relief system at mainline pressure ($5 \times 10^5 Pa$) is practically identical.

Fig. 1.1.7 shows the simplest arrangement for load relief of the slide of a robot

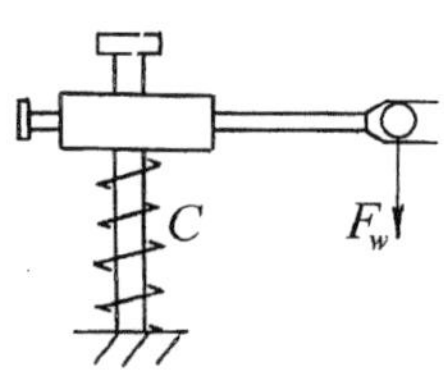

Fig. 1.1.7

arm lift using a compression spring. To prevent over-compensation of the weight at the lowest setting a spring of low stiffness C must be chosen.

In the overwhelming majority of robot kinematic schemes, static load relief must use energy storage with constant or sine-cosine load characteristics [88,99]. The load relief force can be made uniform by using a cam mechanism consisting

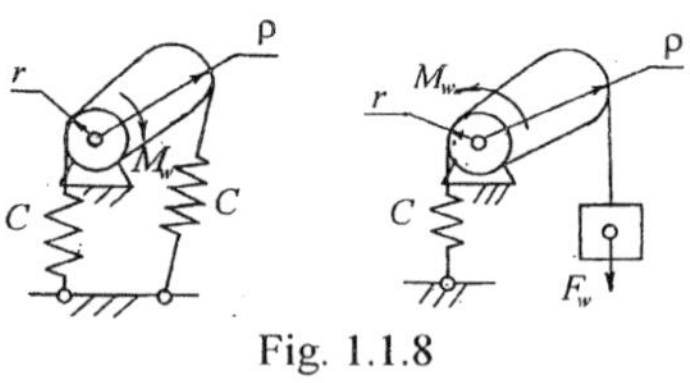

Fig. 1.1.8

of a linear spring and elastic component. The cam profile (Fig. 1.1.8) is chosen from the requirement for uniform torque on its shaft over a particular angle of rotation. The effect of a constant force is obtained by pantograph mechanisms for static load relief of the drive of the robot arm lift (*Patent USSR 1278201, IPC*

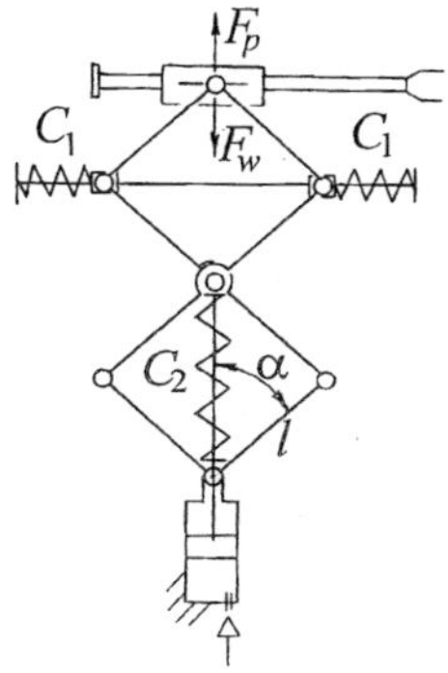

Fig. 1.1.9

B25J9/00). The output characteristic of such a load relief system (Fig. 1.1.9) takes the form

$$F_r = 2l[2C_1 \cos\alpha + C_2(1-\cos\alpha)],$$

where l is the length of the pantograph component; C_1 and C_2 are the stiffnesses of the compression springs; α is the angle between the longitudinal axis and the pantograph components. When $2C_1 = C_2$ the load relief force will be constant at $F_r = 2lC_2 = const$ in the range $\alpha_r = [\varepsilon, 90^\circ - \varepsilon]$ of the working angle where $0 < \varepsilon \ll \alpha_r$, which allows complete load relief of the lifting robot arm drive from the effects of the weight force of the moving components.

Load relievers with sine-cosine load characteristics are used as compensators for the shafts of rotating robot components where the torque M_w from the weight of the components is

$$M_w = mgR\sin\varphi\,. \tag{1.1.5}$$

Here m is the component mass; φ is the rotation angle of the component; R is the offset of the rotation axis from the centre of mass of the component. Much specialist literature deals with the problem of designing load relievers with sine-cosine load characteristics [81, 88, 108]. Different relief characteristics can be produced by a variety of schemes using springs which may be coupled directly to the robot components or operate through additional structural elements such as rollers, cams, links, elastic components or gear sets, etc.

As well as springs, wide use is made of pneumatic systems to compensate torques of the form (1.1.5). Fig. 1.1.10, *a* shows the welding robot *KUKA* in which load relief of the column drive *1* inclination is provided by the two pneumatic cylinders *2*, which are fed by mainline air (pressure $p = 5\times10^5\,Pa$) from the plenum chamber *3*. Using the scheme shown in Fig. 1.1.10, *b*, the equation for the relieving torque M_p can be written in the form

$$M_p = F_p r\sin\alpha = pSra(a^2 + r^2 - 2ar\cos\varphi)^{-1/2}\sin\varphi\,, \tag{1.1.6}$$

where $F_p = const$ is the force developed by the pneumatic cylinders *2*; S is the

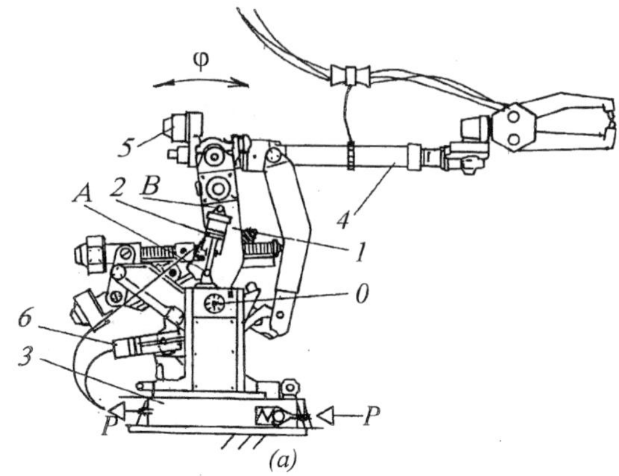

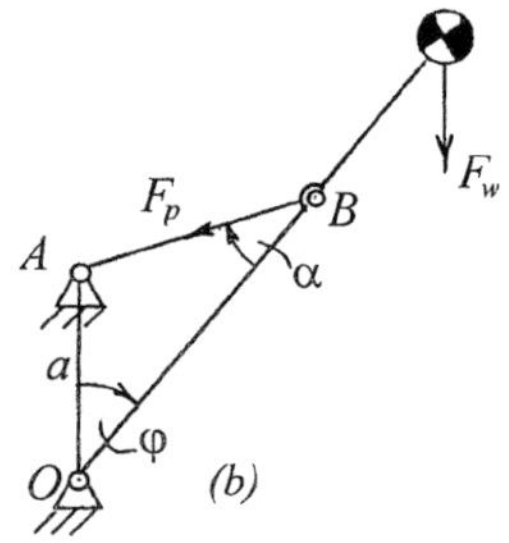

Fig. 1.1.10

area of the piston in the pneumatic cylinder; $r = OB,\ a = OA$. It is easy to see that such a load relief method ensures complete load relief for the drive of the column inclination only at the two points $\pm\varphi_p$ when the condition $M_w = M_p$ is satisfied:

$$2ar\cos\varphi_p = a^2 + r^2 - \left(\frac{arpS}{mgR}\right)^2 \qquad (1.1.7)$$

However, even approximate load relief yields a considerable reduction in energy use and in the loading of the drive of the column inclination.

Static load relief of the drive of the slewing arm *4* is ensured by providing it with partial static balance by the electric motors *5* and load relief from the two pneumatic cylinders *6*.

In robots of the *IR 100 KUKA* series a pneumatic-hydraulic system (Fig. 1.1.5) is used for passive static load relief of the drive of the column *2*. Its accumulator consists of a metallic vessel *3*, in which a rubber diaphragm *4* separates an inert gas *5* and oil *6*. The cavity of the accumulator that is filled with oil is connected by a pipeline to the hydraulic cylinder *7*, whose piston rod acts on the column *2* and partially relieves the inclination drive. The accumulator is filled to a pressure $p = 1.6\times10^7\,Pa$ in the position when the column *2* and arm of the robot are vertical, i.e. in the position of unstable equilibrium.

In addition to improving robot dynamics by using static load relief systems, the mechanical parts of the manipulator and drive controls have increasing demands placed on them. The vibration activity of the whole robot can increase due to the use of elastic elements from which forces that vary significantly are transmitted simultaneously to the driven components of the drive and onto the housing. When considering control of robots, it is necessary to take into account the type of load relief in the drive.

Static load relief of robot drives takes no account of the changing character of the inertia forces. Contemporary industrial robots are machines with high operating speeds, which implies that inertia forces are overcome significantly. These forces are of the same order of magnitude as the weight, but in many cases drives are loaded exclusively by inertia forces, for example when moving in the horizontal plane.

§1.2 Systems to relieve drives from inertia forces

In general, the compensation of inertia forces in universal industrial robots is extremely difficult. Their components have complex transient motion when the acceleration and deceleration phases change continuously at a variety of rates. For this reason, systems for inertia load relief are not widely used in robots with contour and positional control systems. However, in practice every cyclic robotic system has load relief from inertia forces.

The problem of relieving drives from inertia loads arose long before the appearance of robots and is connected with the problem of *reducing vibration activity in continuously operating cyclic machines* such as printing, textile and other such machines [70, 89, 101]. The drives of such machines experience pulses of energy between the driven components and the driver due to the accumulation and recovery of kinetic energy from the non-uniform motion of masses during steady state conditions when energy consumption is practically constant. In a closed kinematic cycle of motion there is no energy increase. Consequently these energy pulses are additional to the average value supplied by the driver and have reversed sign. These additional pulsating forces can often considerably exceed the value of the average driving torque, and can cause transmission overload, collisions between contacting surfaces and noise intensification. Load relief ensures maximum reduction of excess energy in the range where pulsation occurs, and is achieved by introducing an additional load relief arrangement in the drive. This stores the excess energy during the return portion of the actuating mechanism cycle, and releases it during the duty cycle. The storage and release of excess energy must be programmed to ensure that during every phase of the kinematic cycle the excess and load relief torques coincide in value but are opposite in direction.

The simplest means of reducing the vibration activity in continuously operating cyclic drives and load relieving of their components are *inertial dynamic vibration absorbers* [107]. Fig. 1.2.1 shows the arrangement of a spring absorber *1* for torsional vibration of shaft *2*, where J_d, C_d and B_d are the moment of inertia,

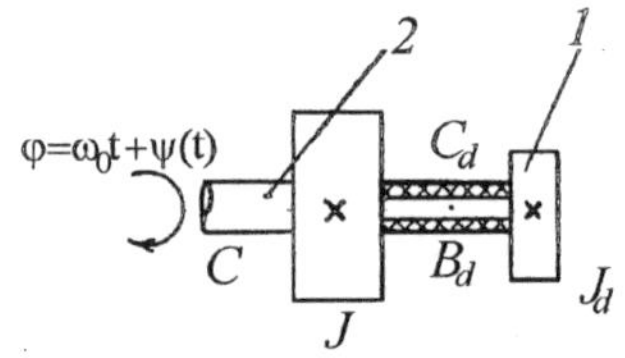

Fig.1.2.1

stiffness and resistance coefficient, respectively, of the dynamic absorber. When the inertial absorber is connected the rotational vibration energy is redistributed from shaft *2* to the absorber *1*. Absorbers of this type are used to relieve components undergoing single harmonic and narrow band random vibration. The relief efficiency is defined as the reduction in the dynamic error $\psi(t)$ in the angle $\varphi(t)$ of shaft *2* of the drive after connecting the absorber. Using a resonant absorber arrangement provides the greatest effect, when the resistance coefficient B_d of the absorber shaft is small and the parameters J_d and C_d are chosen so that the partial frequency of the attached absorber $\omega_d = (C_d / J_d)^{1/2}$ is close to the frequency of vibration of the disturbed shaft. To increase the frequency band in which a absorber with low dissipation is effective, various self-adjusting inertia absorbers are available (rollers, pendulums, smoothly acting weights, spring driven weights), and absorbers with non-linear elastic characteristic or dynamic absorbers of the dissipative type (vibration damper) may be used [107]. The latter type of absorber does not redistribute the vibration energy but only improves vibration energy dissipation.

Active systems for dynamical absorbtion of vibration are used to relieve drive components with unstable operating speeds. These consist of systems that control and supplement the energy sources automatically. Vibration damping for continuously varying vibration loads is accomplished by continuous adjustment of the absorber's dynamic parameters [107].

Passive (Fig. 1.1.3) or combined (Fig. 1.1.4) load relief systems for cyclic drives are based in practice on a potential energy store that consists of elastic elements (either springs or a pneumatic device) and a carrier medium (such as a cam or lever mechanism). The forces developed by such load relief systems are functions of the position of the driven shaft, but do not depend on its velocity. As a result a cyclic drive with a potential energy store becomes in fact the vibrating system. By mutually balancing the elastic forces from the reliever and the inertia forces from the moving components of the drive, the drive can be relieved from inertia forces. It is well known [8] that elastic and inertia forces are mutually in equilibrium for systems at resonance. Hence, the energy of the driver in this case is only used to overcome the dissipative forces and to execute active mechanical work.

In vibration theory, resonance is understood to mean the relatively large selective response of a linear vibrating system due to a periodic influence at a frequency close to its own natural frequency. Due to resonance, there is an abrupt growth in the amplitude of vibration produced by a given excitation force, and conversely, any required amplitude of vibration occurs with minimal excitation force from the excitation vibrator.

In non-linear vibrating systems with sufficiently large nonlinearity, resonance can, in practice, occur with arbitrary periodic excitation. This can be explained by the fact that the vibration frequency depends on the energy (non-isochronous) and on the presence of higher harmonics in the spectrum of periodic vibration (anharmonic vibration of a nonlinear system).

To increase the efficiency of a resonant load relief system it is necessary to: connect the energy accumulator as close as possible to the basic source of the inertia force, to ensure that the resonance tuning is observed strictly; to ensure agreement between the parameters of the drive, the vibrating system and the processing loads [12].

Resonant load relief for a *continuously operating cyclic drive* has particular features. From the known variation of the inertia force, the transfer function is synthesised for a carrier medium (such as a profiled cam or a lever mechanism)

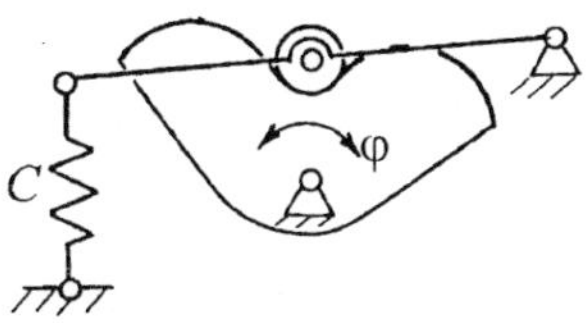

Fig. 1.2.2

which ensures a compensating effect on the driven mass from the relieving force. The frequency of rotation of the driver is usually assumed to be nominal [89]. Fig. 1.2.2 shows an arrangement for a cam load relief device in which the cam profile is attached to the transmission output component and interacts with the elastic elements through a lever and supplies a torque to compensate the pulsating inertia force on the output component. The change of sign in the compensating moment during transfer from the stored energy on return is supplied by arranging that the cam profile normal passes through the centre of rotation of the cam.

Depending on where the load relief mechanism is attached to the transmission of a continuously operating cyclic drive, three basic types of load relief systems can be distinguished [107]: "output component to frame", "input component to frame", "input component to output component".

Load relief using the first arrangement (Fig. 1.2.3, *a*) is the best example for compensating inertia forces on the driven components of a cyclic machine. This arrangement relieves the load on the whole drive [93]. The second type of arrangement is shown in Fig. 1.2.3, *b* and makes it possible to balance

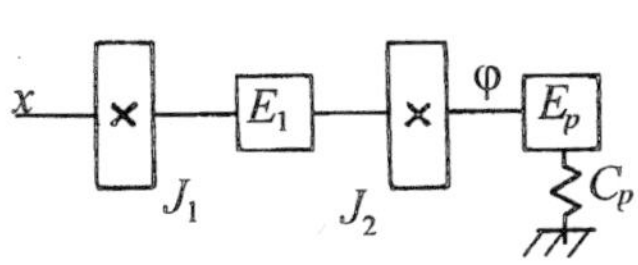

Fig. 1.2.3*a*

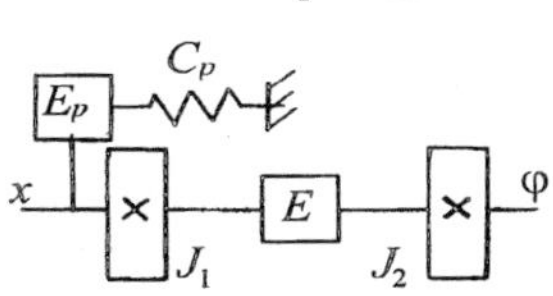

Fig. 1.2.3*b*

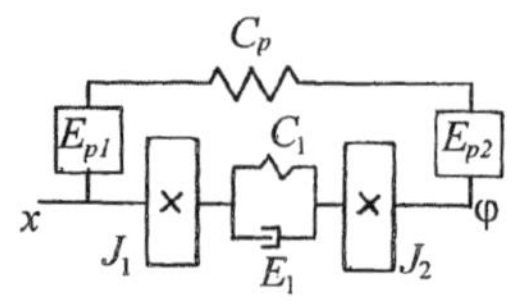

Fig. 1.2.3*c*

disturbances acting on the driver from the transmission. This arrangement reduces only the varying components of the general loads on the drive, but the load on the transmission remains almost unchanged. To reduce the loads passing through the transmission, the load relief arrangement required is of the third type as shown in Fig. 1.2.3, *c*. This arrangement can be used for dynamically loaded transmissions.

In addition to their undoubted merits, load relief systems also have drawbacks and peculiarities. The principal one is that load relief systems transmit large disturbing effects to the frame. For this reason, when using the schemes of Fig. 1.2.3, *a, b* for synthesising load relief, the aim must be to decrease the movement and reduce the structural weight of the load relief mechanism, which makes it possible to eliminate the distorting effect of the inertia loads [107]. It must also be mentioned that balancing the rotating masses of the drive [93] is usually carried out before relieving the load in a continuously operating cyclic machine. The effectiveness of load relief devices with elastic elements becomes less as the operating regime moves away from resonance [89]. In addition, for some operating regimes a load relief system can worsen the dynamic qualities of the machine [93]. It can also be difficult to start a drive having load relief since the driver is loaded by an additional torque equal to that of the inertia load at resonance with no load relief [12].

Methods for compensating inertia forces in continuously operating cyclic drives and in the drives of cyclic robots also have many similar problems. However there are differences caused by the operating regime with dwell time during point-to-point positioning of the driven element of a robot. Additionally, cyclic robots, as shown, have between three and five degrees of freedom. A robot operating element, a prime example of which is a gripper device, undergoes pick-and-place motion during successive operation of the drive for each degree of freedom. The sequence and velocity of operational motions are defined by a cyclic control system, and the displacements are controlled by rigid stops.

Basically, the cyclic robot drives operate in two regimes. The first, which also represents machine filling systems, materials handling and tilting mechanisms, can be characterised as a cyclic reversal operation with dwell-time at the extreme position [111]. During this the motion of the driven components consists of fly-back or swinging motion near the stationary axis.

In another operational configuration, inherent in automatic vehicles, stacking cranes, tracked manipulators, rotating tables and rotating heads, the drive motion has a step-by-step reversing character. Depending on the active program, the cycle of motion of the driven components can include variations through one or more steps.

Both configurations are coupled with frequent stopping and reversal of the drive. During intensive acceleration and braking the transmission undergoes significant pulsating loads due to the inertia forces. As a result, the driving torque developed by the driver is significantly larger than the rotational torque required to move the operating element with constant velocity and the friction forces in the drive increase considerably.

Load relief of cyclic robot drives from the effects of inertia forces can be achieved by a load relief system consisting of absorbers or storage devices for either kinetic or potential mechanical energy.

Since the movements of a cyclic robot along the degree of freedom are consecutive, the dynamics of the whole robot can be described by independent equations of motion for each separate chosen degree of freedom. Later we will consider a generalised model for the separate degrees of freedom of a cyclic robot.

Fig. 1.2.4 shows schematically the arrangement for a robot swivelling arm drive with *flywheel energy storage* [113]. Arm *1* of the robot, is connected through the clutch *2*, mechanical transmission *3*, lever *4* and connecting rod *5*, to a kinetic energy store that consists of lever *6*, reduction transmission *7*, clutches *8* and *9*, flywheel *10* and electric motor *11*, which is used to compensate the dissipative losses in the drive.

The length of components *4*, *5* and *6* is chosen so that as the arm rotates to positions *I*, *II*, *III* and *IV* the transmission ratio between them changes smoothly from 0 to $\pm\infty$. In particular, this occurs when the length of the levers is identical

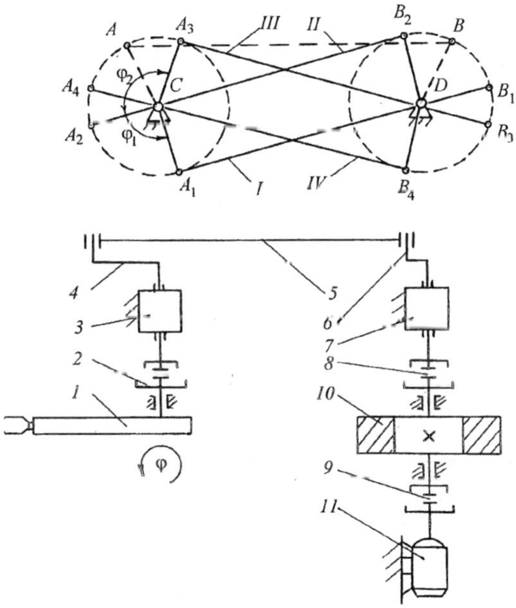

Fig. 1.2.4

and the length of the connecting rod is such that when one lever is perpendicular to the connecting rod the other lies along it.

The drive operates in the following manner. When arm *1* of the robot is placed at the end position, the clutches *2* and *8* are disengaged with the four-bar linkage stopped in position *I* so that the angles $B_1A_1C = 90°$, $A_1DB_1 = 180°$. As a result, flywheel *10* rotates autonomously and is "trickle-charged" with kinetic energy from the low power motor *11* through clutch *9*. To operate arm *1* clutches *2* and *8* are engaged while clutch *9* is disengaged. The rotating flywheel *10* is connected to the stationary arm without impact, since the transmission ratio from lever *4* to lever *6* is equal to $-\infty$. When lever *4* rotates from position *I* to position *II* through an angle $\varphi_1 = 90°$, the robot arm is accelerated by flywheel *10* and automatically changes the transmission ratio between the levers from $-\infty$ to 0, which ensures that the flywheel is completely at rest in position *II*. Subsequently, at angle φ_2 the robot arm is braked and held in position *III*. As a result, the flywheel is accelerated in the opposite direction to its maximum angular velocity. After this, clutches *2* and *8* are disengaged. In this way the rotation of the lever *4* from position *I* to position *III* is ensured while a large part of the energy is preserved in the flywheel, which might be "trickle-charged" as required by the reversing motor *11*. To ensure rotation of the robot arm to the starting position clutches *2* and *8* are engaged and *9* disengaged. As the four-bar linkage moves to position *IV* the robot arm is accelerated. The flywheel will stop when the rotation angle reaches φ_2. After this the mechanism returns to its starting position. Energy losses during rotation of the robot arm are connected only with dissipation due to resistance in the drive. It is important that the rotation angle of the robot arm should be accurately manipulated automatically by selecting the relationship between the lengths of the four-bar linkage elements and depends on the dynamics of rotation only through the elasticity of the transmission elements. When they are sufficiently stiff, then the stopping accuracy is not affected when the masses (moments of inertia) of the driven robot components and their velocities are all increased.

The use of flywheels in energy storage systems for load relief allows a reduction in energy usage and an increase in performance [49, 50]. Interfacing flywheels with variators or with lever transmission mechanisms [51] makes it possible to implement movements with smooth accelerating and braking of the operating element [56] which is important for the accuracy and endurance of robots.

Damping - the most widely used method of load relief, is used to reduce velocity and cancel impact of moving components as they approach end points during point-to-point movements. Fig. 1.2.5 shows schematically a double-acting hydraulic absorber. The hydraulic cylinder *1* with a double acting piston is filled with oil. When the piston rod moves along the x – axis the piston expels oil from the working cavity through the throttling system *2*. The compensator *4* and reversing valve *3* serve to compensate for leakage from the hydraulic cylinder. Apart from hydraulic damping, friction, pneumatic or electro-magnetic braking are widely used to relieve drives from inertia forces at the end points of the motion [40]. Damping, which is basically energy dissipation, does not reduce the energy consumption of drives. Moreover, hydraulic damping requires frequent adjustment of the throttling system due to temperature changes (and consequently viscosity)

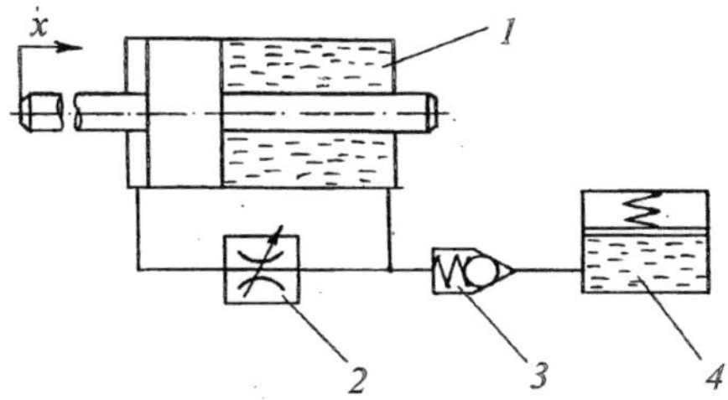

Fig. 1.2.5

in the oil during operation.

The application in cyclic robotic systems of potential energy accumulators consisting of elastic elements makes it possible to organise the work of the robot in the form of high efficiency resonant motions of the operating element. Resonant operational regimes are characterised by low energy requirements and high robot operating speeds under intensive acceleration and non-impact braking.

A *resonant robotic system (resonant robot)* [4]is a cyclic industrial robot with potential energy accumulators consisting of elastic elements whose motion is performed as resonant mechanical vibration of the operating element with dwell times at the positioning points. Correspondingly, a *resonant drive* will be taken to refer to the drive of an isolated degree of freedom for a resonant robotic system.

The structure of a resonant drive can be formed either using passive (Fig. 1.1.3) or combined (Fig. 1.1.4) load relief of the drive from the effect of the inertia loads due to moving masses.

In passive load relief systems (Fig. 1.1.3) the element to be controlled is the operating component *OC*, which can be a gripping device or a component carrying one of the degrees of freedom of the robot. The energy store ES, consisting of a potential energy accumulator, develops the basic motion forces in the drive, which balance the inertia forces of the moving parts of the transmission *TR* and the operating component. The driver *D1* performs work to compensate the energy losses in the drive. To hold the operating component at positioning points the regulating component *RC* is used. The drive operation is coordinated by the robot control system *CS* using the feedback system FS, which records the status of the elements and the phase coordinates φ and $\dot{\varphi}$ of the operating component.

A resonant drive operates as follows: from the initial position, the mechanical energy store is charged, the driver *D1* is de-energised and the operating component is held fixed. On command from the control system, control voltages U_1 and U_3 are delivered to the driver *D1* and the holding device respectively. As a result, the transmission is released, the operating component is extensively accelerated by the driving forces from both the accumulator elastic elements and the actuator until the point where the accumulator is not completely discharged. Then, depending on the required displacement, the accumulator can either be disengaged from the transmission at some portion of the path (*Patent USSR 1399110, IPC B25J9/00*), or start to charge, damping the kinetic energy. As it approaches the required positioning point, the operating component is moving with low velocity, and the accumulator is fully charged. At the required positional point, the operating component becomes fixed. The driver control method during motion depends on the required motion parameters in the presence of positive compensation of the dissipation losses in the system.

Passive load relief of drives provides the basis for the construction of a *cyclic resonant module* [5], which is shown schematically in Fig. 1.2.6. The module

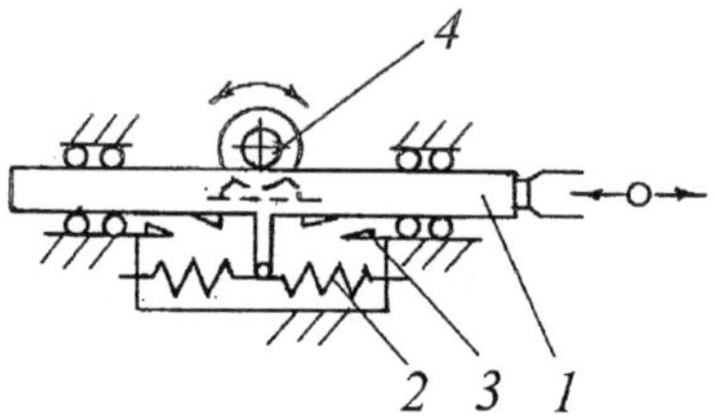

Fig. 1.2.6

consists of a vibrating system with carriage *1*, attached to the frame by spring *2*, which acts as a potential energy accumulator. At each end of the carriage stroke, latches *3* are arranged which prevent motion of the carriage in the reverse direction after stopping. The carriage is connected with a reversible electric motor *4* controlled in terms of position and velocity of motion of the carriage by a link of the feedback system. The module works in a resonant vibration regime [12], with smooth acceleration and gentle braking of the carriage. When the module is tuned to resonance, forces developed by the motor are only used to compensate energy losses in the drive and consist of useful work. Forces supplied by the accumulator balance inertia forces [5, 7].

The resonant robot *MARS* [4] consists of three resonant modules, two of which deal with sliding and the third with rotation. Fig. 1.2.7 shows its kinematic arrangement and design. The parameters of the *MARS* robot compared to those of the cyclic robot *Speedy One* (Italy) are given in the following table:

Principal attributes	*MARS*	*Speedy One*
Configuration	Modular	Modular
Control system	Cyclic	Cyclic
Maximum load capacity, kg	0.5	1.5
Arm displacement, mm		
- along x-axis	210	280
- along z-axis	100	180
Rotation about z-axis, degrees	120	180
Positioning accuracy, mm	0.05	0.1
Velocity		
- rotation about z-axis, deg/s	300	200
- along x-axis, m/s	2.0	1.0
- along z-axis, m/s	2.0	2.0
Overall dimensions, mm	575x400x150	400x650x800
Mass, kg	8.0	30.0
Operating power, W	20.0	1000.0

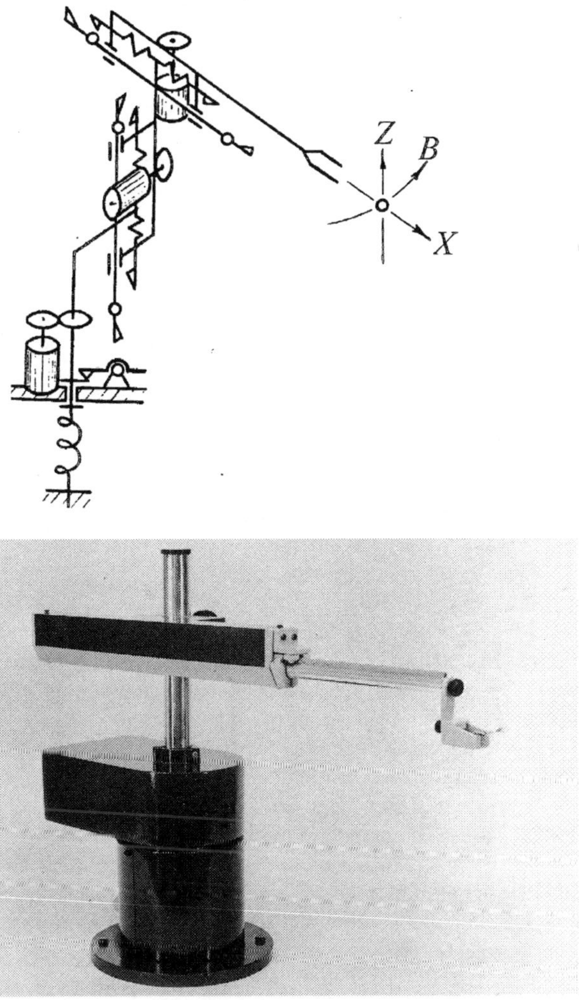

Fig. 1.2.7

The resonance tuning of the drives for the *MARS* robot ensures minimum electrical energy expenditure and a high operating speed.

Stepped motion of the operating component under resonant conditions is achieved by using a *stepped resonant drive* (*Patent USSR 1357219, IPC B25J9/00*). In this (Fig. 1.2.8), the operating component *1* is kinematically connected to the energy accumulator - spring *2*, the controlling latch *3* and the driver *4*. Connecting one end of the spring to the crank *5* ensures unstable equilibrium positions of the crank during motion of the operating component. Under the action of the torque from the driver and the spring, the crank rotates and each rotation causes one step. After the required number of steps of the operating

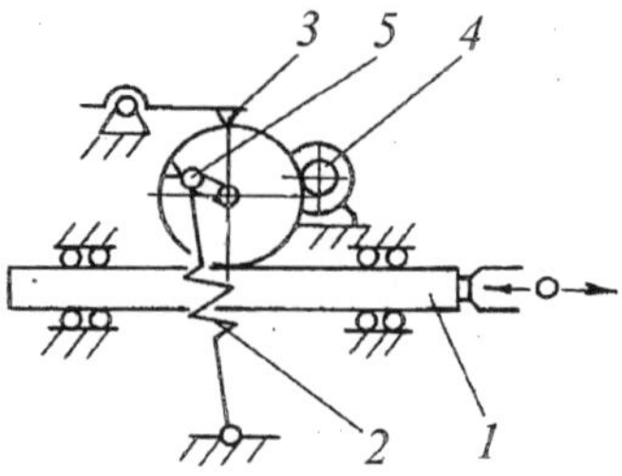

Fig. 1.2.8

component, the latch *3* engages. In the fixed position, the ends of spring *2* are at maximum opening and lie in a straight line passing through the axis of crank 5.

The increase of energy consumption of a stepped resonant drive above the energy used by a cyclic resonant drive is explained by the additional energy expenditure necessary to disturb the stepped drive from the position of unstable equilibrium during the initial phase of motion. On the other hand, fixing the drive in an unstable equilibrium position makes it possible to ensure that the drive for the fixing mechanism *3* is considerably lighter. The best ideal possibilities for cyclic and stepped resonant drives are considered in Chapter 3.

Cyclic robot drives can be relieved from load by using combined load relief systems (Fig. 1.1.4). Fig. 1.2.9 shows one of the possible methods of achieving this in practice (*Patent USSR 1453108, IPC F16H33/02*). The drive consists of a basic driver *1*, transmission *3*, transmission mechanism *5*, clutch *6*, controlled brake *7*, potential energy accumulator *8*, in the form of torsional spring *16*, self-locking reducer *9*, supplementary driver *10* and two position sensors *11* and *12*. Before initial acceleration accumulator *8* is charged using driver *10*. The end of the spring is restrained during this process by brake *7*. The degree to which the accumulator is charged is controlled by the position sensors *11* and *12*, attached to the ends of the spring. Motion of the drive is effected by the applying stored energy from the accumulator to the operating component *4* through clutch *6* by means of the transfer mechanism *5*. Clutch *6* is engaged simultaneously with the

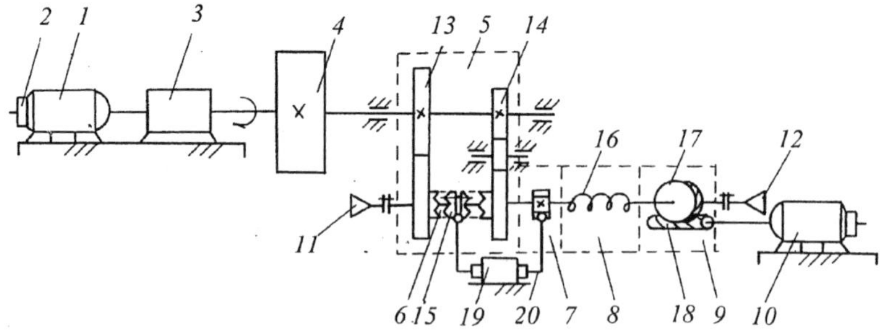

Fig. 1.2.9

release of brake *7* by means of the lever 20, on which acts electromagnet *19*. The potential energy stored by the spring is converted to kinetic energy of the accelerated drive. As the operating component reaches its programmed velocity as measured by the tachogenerator 2, clutch *6* disengages and brake *7* is engaged.

For braking, the accumulator is connected by clutch *6* to the operating component of the drive. The kinetic energy of the moving components is transformed into potential energy by twisting the spring. At the instant of stopping, clutch *6* is disengaged and brake *7* is engaged. The accumulator is securely maintained in the charged condition.

It is possible to generate complex motion for cyclic robots with increasing and decreasing velocities and to relieve the drive completely from inertia load effects. Using such a load relief scheme it is possible to reduce the mass of the basic drive and to reduce the power supply to the driver. The latter is important for automatically guided and robotic vehicles with autonomous energy sources.

As can be seen, there is an extensive range of methods for load relief in cyclic robot drives. Every load relief method has its advantages and disadvantages. In every concrete case the choice of one or other load relief system is based on many factors. The basic factors are: the form of motion of the operating element (cyclic or stepped); the type of principal drive; the speed, energy consumption, material requirements and vibration activity of the drive. It is clear also that the dynamic qualities of cyclic robot drives can be improved by using load relief systems that are based, in particular, on the high efficiency of resonant processes. The performance of resonant robotic systems is determined in principal by the properties of the potential energy accumulator. The energy consumption of a driver becomes minimal even in drives without a damping arrangement and braking of the operating component occurs smoothly.

We shall look at several possible implementations and at the characteristics of different components of resonant robotic systems. We shall limit the discussion to only short examples, since detailed descriptions can be found in specialist literature.

§1.3 Resonant robotic system components

The study of standard commercially used components will be carried out on the basis of their use in resonant robotic systems and the corresponding mathematical description.

1.3.1. The basic component in the mechanical parts of a resonant drive is the **transmission**, consisting of mechanisms, each of which from the operational point of view can be considered as a mechanical system with holonomous stationary constraints having coordinate conversions in the form

$$q_s = \Pi_s(x), \quad s = 1, \ldots, m, \tag{1.3.1}$$

where x is the generalised coordinate of the input component, $q_1, \ldots, q_m$ are the output coordinates (in generalised form), $\Pi_1, \ldots, \Pi_m$ are scalar functions of mechanism position [111].

The choice of dynamic model for the transmission depends on the complexity of the problem studied and the necessity to obtain a visible solution. Thus in the preliminary stage of dynamic synthesis a kinetostatic model of the transmission with rigid components is used, i.e. a *rigid machine* [12]. The principal assumption in the model is that all transmission elements (gearing, shafts, bearings and operating components) are rigid bodies. This model makes it possible to obtain the initial information about the level of the dynamic loading of the transmission, to find the forces necessary to produce the required motion, to calculate the reactions in the kinematic pairs [112]. Although this is the simplest dynamic model it represents the main dedicated function of the transmission, which is to transform the motion according to the given program [66].

The motion of the rigid transmission is described using equations in the form of Lagrangian equations of the second order [66, 106]

$$\frac{d}{dt}\frac{\partial T}{\partial \dot{q}_s}-\frac{\partial T}{\partial q_s}=Q_s,\quad s=1,\ \dots,\ m\,, \tag{1.3.2}$$

where T is the total kinetic energy of the moving components of the transmission, q_s are generalised coordinates, and Q_s are generalised forces. The generalised coordinates q_s are degrees of freedom that can be varied independently and that are chosen by their convenience for specifying the potential and kinetic energy.

In the study of rapid acting highly loaded robots more complex models are used to take account of the dissipative peculiarities of the flexible transmission and the inertial features of the flexible driven component [70,112]. Additional generalised coordinates $\theta_1, \dots, \theta_n$ must be introduced to reflect the component deformations as a result of which the position function takes the form [106] $q_s=\Phi_s(x,\ \theta_1,\ \dots,\ \theta_n),\ \ s=1,\ \dots,\ m+n$. In such models the number of degrees of freedom is larger than the number of degrees of mobility and the model is called *a flexible robot* [67].

1.3.2. Drive actuators. *Independently excited constant current electric motors (DC torque motors)* are widely used in resonant robot systems because of their high dynamic qualities and simple control systems [12, 37, 106]. The choice of dynamic model for the actuator depends on the velocity at which the processes occur in the drive, the frequency spectrum of the forces acting and the complexity of the solution sought. Within the wide range of changes in the input and output variables, the dynamic characteristics of a constant current independently excited electric motor with controlled armature circuit takes the form [12, 37, 67, 106]

$$u=E+Ri+L\dot{i}\,, \tag{1.3.3}$$

where u is the controlled voltage in the armature; i,R,L are respectively the current, active resistance and inductance of the armature windings; E is the back emf *(electromotive force)* generated as the rotor turns, which can be found from the relationship

$$E=\frac{hN}{2\pi a}\Phi\dot{x}=k_e\dot{x}\,. \tag{1.3.4}$$

Here h is the number of pole pieces in the motor; N is the number of active conductors; a is the number of parallel branches in the armature windings; Φ is the magnetic flux in the excitation coils; $\dot{x}$ is the rotational speed of the rotor. The driving torque Q is proportional to the armature current

$$Q = \frac{hN}{2\pi a}\Phi i = k_m i\,. \tag{1.3.5}$$

Substituting (1.3.4) and (1.3.5) into (1.3.3) we can find

$$\tau_e \dot{Q} + Q = k_m R^{-1} u + k_m k_e R^{-1} \dot{x}\,, \tag{1.3.6}$$

where $\tau_e = LR^{-1}$ is the electromagnetic time constant. Note that in the SI system the coefficients $k_m = k_e = k$ [92].

The motor dynamic characteristic (1.3.6) is used during the study of fast acting drives in which the velocity at which the processes occur in the transmission at a speed comparable to that at which the electromagnetic processes occur in the motor [12, 37, 112]. In cases when the load variation in the transmission or the velocity of motion is quite slow, then we may put $\tau_e = 0$ in (1.3.6), and use the motor static characteristic, which has the form [12,66]

$$Q = kR^{-1}(u - k\dot{x})\,. \tag{1.3.7}$$

1.3.3. Feedback transducers. The characteristics of the transducers used as the sensitive elements of the feedback system of a robot drive must satisfy the requirements of maximum sensitivity and minimum inertia, be stable and also have the linearity and uniqueness of the static characteristics [72].

a) Tachogenerators. As a rule, both independently excited constant current tachogenerators and asynchronous tachogenerators can be used as velocity sensors in robotic systems with electric motors (Fig. 1.3.1) [104]. It is also possible to use differential transformers as sensors.

A tachogenerator is either a constant current or asynchronous electrical machine whose shaft is rigidly attached to the output shaft of the motor whose angular velocity is measured.

The transfer function describing relations between the angle of rotation of the tachogenerator and the output voltage takes the form of the transfer function for an aperiodic element

$$W_{tg} = \frac{k_{tg}}{1 + \tau_{tg} p}, \qquad k_{tg} = \frac{C_e \Phi}{1 + R_a / R_i}\,. \tag{1.3.8}$$

Here, k_{tg} is the transfer coefficient of the generator, R_a and R_i are the active resistances of the armature and load, $\tau_{tg} = L_a (R_a + R_i)^{-1}$ is the time constant, L_a

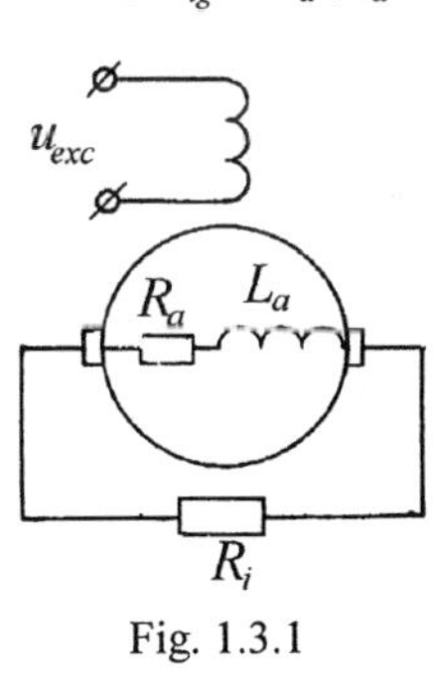

Fig. 1.3.1

is the armature inductance, p is the complex variable (Laplace variable), Φ is the excitation flux.

In a constant current tachogenerator there is an insensitive zone in regions of low angular velocity due to voltage losses in the brushes. Sensitivity lies in the range of $2 \sim 20 mV$ /rpm. The accuracy of measurement of the rotation frequency is between $0.5 \sim 2.5\%$, while for a precision tachogenerator it can be between $0.05 \sim 0.1\%$.

Potentiometric, inductive, encoded or impulse sensors can be used for angular or linear displacements in practice for resonant robots.

b) Potentiometric sensors transform an angular (or linear) displacement α into

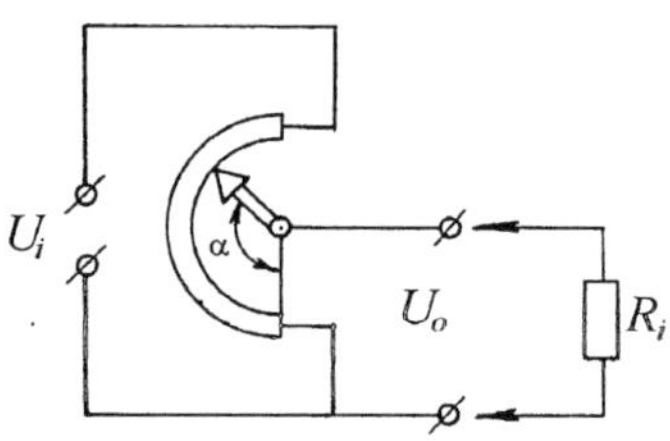

Fig.1.3.2

an output voltage U_o (Fig. 1.3.2). The input voltage U_i can be either constant or variable. Either wire-wound or solid-state potentiometers are used. The relationship $U_o = f(\alpha)$ is nonlinear; the nonlinearity of these sensors is between $0.1 \sim 3.0\%$. The sensitivity of a potentiometric sensor is $\eta = dU_o / dy =$ $= (3 \sim 5).10^{-3} V/mm$, where y is the linear displacement (mm) of the slider corresponding to the rotation angle α.

The basic deficiency of a wire-wound sensor is the stepped characteristic $R(y)$ due to the discrete change in resistance as the slider is displaced. The accuracy of such sensors can reach 0.05%. The accuracy of solid-state potentiometers can reach 0.01%. In calculations the value of the time constant of potentiometric sensors is usually ignored, i.e. the inductance of rheostat windings is completely insignificant.

c) Rotary transformers are alternating current electrical machines with smooth poles. Sine-cosine rotary transformers (SCRT) are used to measure angles. The moving component of the robot mechanical system whose relative rotation must be measured is connected to the stator or rotor of the SCRT. Within the rotor and the stator there are two mutually perpendicular windings (Fig. 1.3.3). When a variable voltage U_1 is fed to the primary windings W_1' of the SCRT, emfs E' and E'' are induced in the secondary windings W_2' and W_2''. These emfs vary according to the laws

$$E' = k_t U_1 \sin\alpha, \quad E'' = k_t U_1 \cos\alpha, \tag{1.3.9}$$

where $k_t = W_1'/W_2' = W_1''/W_2''$ is the transformer coefficient. The output signals of the sensor are proportional to the sine and cosine of the angle of rotation. The

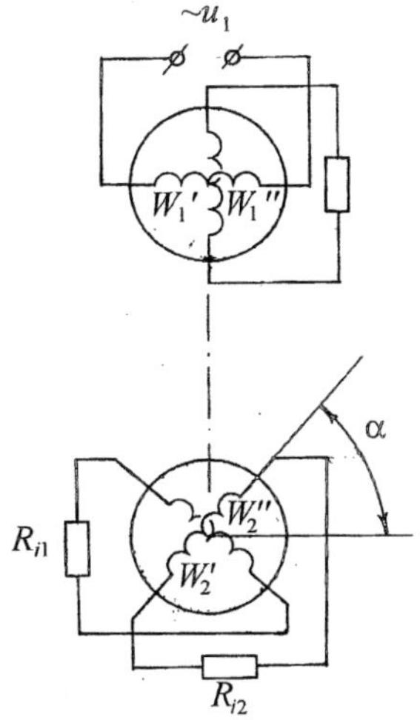

Fig. 1.3.3

angle α is found by computer using the system equations.

For small angles of rotation $(\sim 5^\circ)$, the static characteristic of a rotary transformer can be considered as linear to an accuracy of 0.1%. Special design arrangements exist [31] which make it possible with an accuracy of 0.1% to increase the linear part of the rotation by up to $\pm 55^\circ$.

The sensitivity of rotary transformers is about $1V/\deg$. SCRTs have low errors of $\pm(2 \sim 3)'$ when used to measure angles. Rotary transformers have several disadvantages, which include their high cost and low noise immunity.

d) Encoding sensors transform the shaft displacement into digital code. The sensor consists of a photoelectric (optical) system for sensing together with an encoding disc with stripes whose number is equal to the order of the digits in the output signal. The stripes bear alternately transparent and opaque sectors to imitate 1 and 0 as binary codes. The disc is placed between light sources and photoelectric elements (such as photo-resistors or photo-diodes). The set of signals from the photosensitive elements forms a binary number, which corresponds uniquely with the required angle of rotation of the disc. To eliminate ambiguity, a Gray code mask is superimposed on the disc. As a result, errors are reduced to low order digits. The response speed of encoding sensors is limited only by the velocity of signal transmission to the control system along the information bus.

The drawbacks include high cost, complexity and clearances required for the encoding sensors.

When adaptive control is used in resonant robots, information about the acceleration of the operating elements is often required. Such information can be obtained by differentiating the tachogenerator or position sensor signals. However, it is often easier to use acceleration sensors.

e) Accelerometers used to measure absolute acceleration are basically seismic sensors. Originally a wide variety of construction forms were used but recently the commonest form is the piezoelectric accelerometer (Fig. 1.3.4). This is due to the

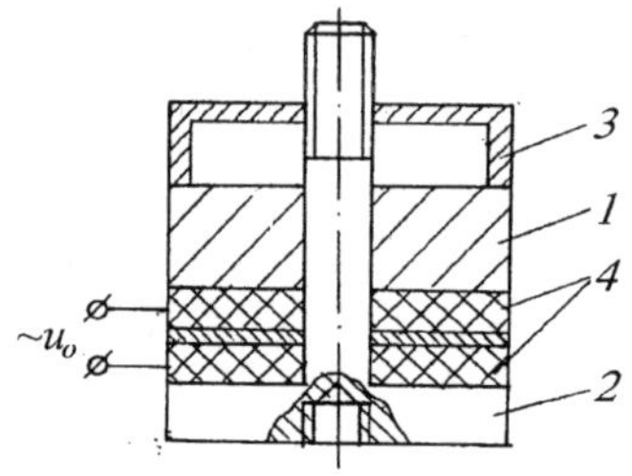

Fig. 1.3.4

wide range of vibration measurement possible $(0.5 \sim 10^4)Hz$, the linear characteristic of the sensor over a wide dynamic range (up to $10^5 \, m/s^2$), their low cost and durability. In addition they do not require an electrical energy source.

As the velocity varies, the "seismic mass" *1* (Fig. 1.3.4) moves relative to the body *2* to which it is connected by spring *3*. As a result the deformation of the piezo-discs 4 changes and a variable voltage U_o is generated at the sensor output.

Choosing the type of accelerometer can ensure the mass of the sensor is kept small compared to the mass of the test object. In some cases, resistance strain gauge, capacitance or inductive accelerometers are used [107].

The use of mechanical energy storage devices in the structure of resonant robots makes it necessary to use force and strain measurement.

f) Force and mechanical strain measurement. For this purpose, resistance strain gauges are the commonest devices used. In dynamic calculations they are considered as inertia-free linear transducers.

To provide a cyclic character of operation in resonant systems is not possible without using contacting or non-contacting limit switches.

g) Limit switches are used to sense the position of the drive mechanism and

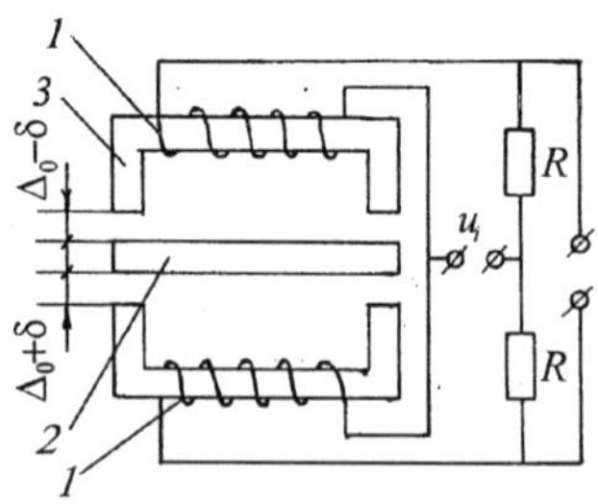

Fig. 1.3.5

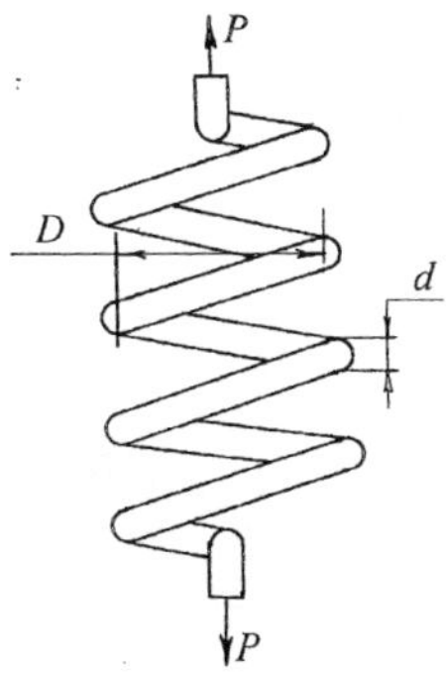

Fig. 1.3.6

also the start, final and intermediate positions as the operating components move. Signals from them are used as input information for the control system to disable the driver and to start the subsequent motion.

Fig. 1.3.5 shows schematically a non-contacting inductive limit switch. At the input coils *1* a variable voltage U_i is applied. Displacement of the bar *2* changes the value of the gaps $\Delta_0 - \delta$ and $\Delta_0 + \delta$, and as a result the magnetic resistance of the circuit, the inductance of the coils, the amplitude of the variable voltage and the output voltage U_o. Assuming that the active resistances of the coils are negligibly small compared to their inductive resistances, and the magnetic resistance of the core *3* is negligibly small compared to the resistance of the air gaps between the bar and core, the sensor characteristic can be considered as linear $U_o = k_s\delta$. Inductive sensors are usually used for small displacements. In robotic systems, they are used in robot positioning zones. As the operating component moves to the target point, the bar enters the gap in the core and the instrument sensor initiates control of the operating component position.

Principally the type of elastic element determines the character of motion of resonant robotic systems and this also determines the form of locking devices.

1.3.4. Elastic elements for storing potential energy.

a) Springs are the most widespread form of elastic element in resonant drives. The returning force of a spring depends on its position and is found from its deformation from the initial state.

For a helical spring when the diameter of the coil *D* is considerably larger than the diameter *d* of the circular stock from which it is wound (Fig. 1.3.6) the elastic force *P* (tension or compression) is directly proportional to the deformation *x*

$$P(x) = Cx. \tag{1.3.10}$$

For a helical compression spring compressed by an initial amount $x_0 = P_0 / C$, where P_0 is the force in the spring in its initial position, *C* is the axial stiffness coefficient, we find

$$P(x) = C(x_0 + x) = P_0 + Cx. \tag{1.3.11}$$

The elastic characteristic for a helical extension spring with an initial length x_l

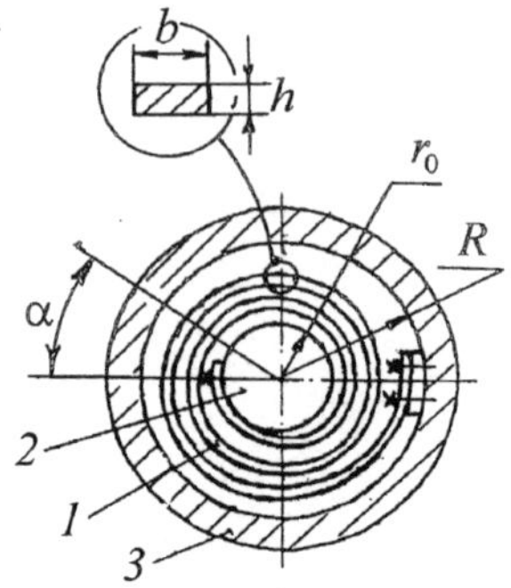

Fig. 1.3.7

takes the form

$$P(x) = C(x - x_t) \tag{1.3.12}$$

The value of x_t is found as follows. Assume that under the action of force P_1 all the coils of the spring are open. The distance between the ends of the spring is x_1. We shall call P_2 and x_2 the force and length of the spring approximately corresponding to the operating deformation of the spring. Then $C = (P_2 - P_1)/(x_2 - x_1)$ and $x_t = x_2 - P_2 / C$. The initial length can be a positive, zero or negative value. The latter case corresponds to a spring with initial tension, when the spring coils in the initial condition are closely coiled against each other, and a force $P_t = Cx_t$ is needed to separate them.

Within the limits of Hooke's law, the axial stiffness coefficient C (*N/m*) is found from the expression

$$C = Gd^4 / 8nD^3, \tag{1.3.13}$$

where G is the shear modulus (for steel springs $G = (7.85 \sim 8).10^{10}\ N/m^2$); n is the number of working coils in the spring.

The potential energy $L(x)$ stored by the spring during deformation, is equal to the work done during deformation by the external force $P(x)$

$$L(x) = \int_0^x P(x)dx\,. \tag{1.3.14}$$

To relieve the load on drives with rotating components, spiral springs with rectangular cross-sections are generally used (Fig. 1.3.7). If a spiral spring is used in pure bending, then the bending torque will be

$$M(\alpha) = \frac{EI}{l}\alpha = C\alpha, \qquad C = \frac{Ebh^3}{12l} \tag{1.3.15}$$

Attaching the spring to the moving component [14]

$$M(\alpha) = \frac{\pi C_k Ebh^3 (n_2 - n - n_c)}{6l}, \qquad 0 \le n \le n_t \tag{1.3.16}$$

Here EI is the spring bending stiffness; E is Young's modulus; I is the second moment of area of the spring cross-section; b and h are to the width and height of the spring cross-section and l is the total length of the coiled spring; $\alpha = 2\pi n$ is the angle of rotation of the spring cross-section; $n_2 = n_t + n_1$ is the number of turns of the spring in the loaded condition; n_t is the number of rotations of the end of the spring; n_1 is the number of turns of the spring in the barrel when released; n_c is the number of turns of the spring in the free state outside the barrel (usually $n_c / n_2 = 0.3 \sim 0.4$); C_k is the spring energy rebound coefficient, whose numerical value depends on the method of attaching the end of the spring *1* to the moving components shaft *2* and barrel *3* (Fig. 1.3.7). The usual methods of attachment are pinned fastening, when $C_k = 0.72 \sim 0.78$; or hinged fastening, when $C_k = 0.65 \sim 0.70$ [46].

The moving component will contain the highest amplitude (potential energy), if the internal radius is r_1 of the unloaded spring inside the barrel is equal to the external radius r_2 of the loaded spring

$$r_1 = r_2 = [0{,}5(R^2 + r_0^2)]^{1/2}. \tag{1.3.17}$$

Here R and r_0 are the internal radius of the barrel and the radius of the internal shaft respectively. The approximate length of the spring will then be $l = \pi(R + r_1)n_1 + 2\pi r_1$.

The potential energy stored by a spiral spring can be found from the formula

$$L(\alpha) = C\alpha^2 / 2. \tag{1.3.18}$$

If two springs are arranged in parallel in the energy storage device (Fig. 1.3.8,

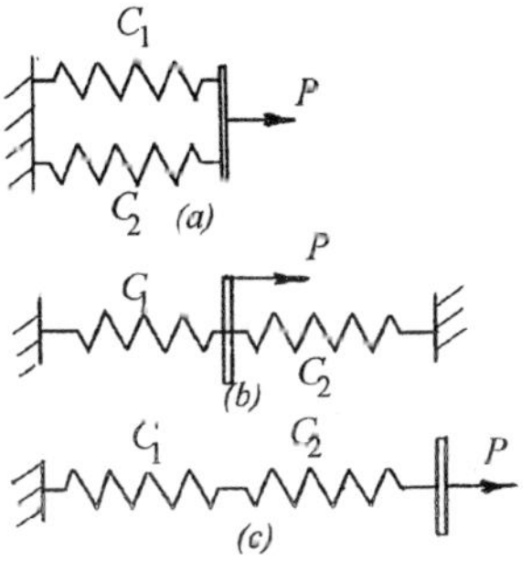

Fig. 1.3.8

a, b) then the total stiffness will be equal to the sum of the individual stiffnesses but the deformation will be $x = P/(C_1 + C_2)$. When the springs are arranged in series (Fig. 1.3.8, *c*) the overall deformation consists of a value equal to the sum of the individual deformations of each spring $x = x_1 + x_2 = P(C_1^{-1} + C_2^{-1})$. The total stiffness in this case will be $C = C_1C_2(C_1 + C_2)^{-1}$.

To produce special forms of motion the components in resonant drives use spring elements with nonlinear elastic characteristics.

Fig. 1.3.9 shows a conical helical spring with a nonlinear load characteristic

[100]. Projecting the form of a spring with n coils onto the horizontal plane we find the equation for an Archimedian spiral $\rho = r_1 + (r_2 - r_1)\theta / 2\pi n$.

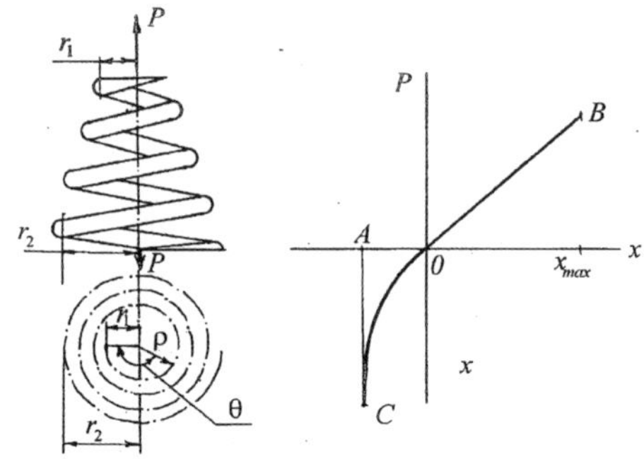

Fig. 1.3.9

Initially the spring is undeformed and its coils are close to one another. We shall find the expression for the axial force P in terms of extension x of the spring. Assuming that the spring coils only undergo torsion it is possible to write down the twisting moment or torque M_t for an arbitrary point defined by the angle θ in the form $M_t = P\rho$. The potential energy of deformation for an element $ds = \rho d\theta$ along the axis of the spring will be

$$dL = \frac{M_t^2}{2GI_p} ds = \frac{16P^2}{G\pi d^4}\left[r_1 + \frac{(r_2 - r_1)\theta}{2\pi n}\right]^3 d\theta$$

Here d is the diameter of the wire from which the spring is wound; G is the shear modulus; $I_p = \pi d^4 / 32$ is the polar moment of inertia of the circular cross-section of the wire.

The extension load P is found from the condition that the work of the force P when moving over distance x must be equal to the potential energy L of the whole spring

$$\frac{Px}{2} = \int_0^{2\pi n} \frac{16P^2}{G\pi d^4}\left[r_1 + \frac{(r_2 - r_1)\theta}{2\pi n}\right]^3 d\theta,$$

from which

$$P = Cx = \frac{\pi G d^4 x}{16(r_1^2 + r_2^2)l}, \tag{1.3.19}$$

where $l = (r_1 + r_2)\pi n$ is the length of the spring. When $r_1 = r_2 = 0.5D$ the stiffness coefficients from equations (1.3.13) and (1.3.19) coincide.

The elastic characteristic (1.3.19) for the extension of a conical spring is a linear function of the deformation x and takes the form of the line OB (Fig. 1.3.9). However, in compression the loaded characteristic of a conical coiled spring is nonlinear and can not be described by equation (1.3.19). From (1.3.19) it can be seen that the stiffness of a coil spring increases as the number of coils and the spring diameter are reduced. Consequently, the lower coils of a conical spring are softer compared to the stiffness of the upper coils. In compression the lower coils are squashed onto one plane and stop working. The number of working coils reduces while the spring stiffness increases. The loaded compression characteristic takes the form of curve OC in Fig. 1.3.9. This "nesting" effect of the coils in a conical spring is used in resonant drives of compact design when the operating components have significant displacement.

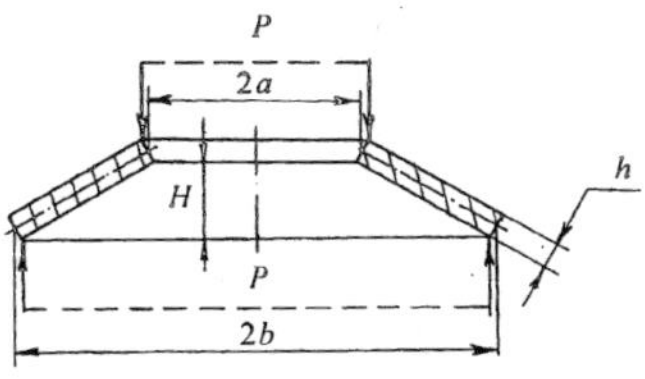

Fig. 1.3.10

The nonlinear characteristic under load for a *disc washer (Belleville washer)* type of spring takes the form [107]

$$P = \frac{\pi E h}{6(b-a)^2} x\left[(H - x/2)(H - x) + h^2\right]\ln\frac{b}{a}. \qquad (1.3.20)$$

Here E is the modulus of elasticity; a, b, h, H are geometric parameters of the spring as shown in Fig. 1.3.10. Disc springs of this type have high load capacity and considerable structural damping between the separate discs, when arranged in parallel or series.

To obtain high torques M for small angles of rotation α, torsion elements are used [89]. The characteristics under load for a torsion element as shown in Fig. 1.3.11 take the form

$$\begin{aligned} M_a &= \frac{\pi G d^4}{32 l}\alpha, \\ M_b &= \frac{\pi G}{32 l}(d_2^4 - d_1^4)\alpha, \\ M_c &= \frac{\pi G}{32}\left[\frac{l_1}{d^4} + \frac{l_2}{D_2^4 - D_1^4}\right]^{-1}\alpha, \\ M_d &= \frac{G b^3 h}{3l}\left[1 - 0.6\frac{l}{b}\right] n\alpha. \end{aligned} \qquad (1.3.21)$$

Here G is the shear modulus; n is the number of plates in the assembly; $d, d_1, d_2, D_1, D_2, l, b, h$ are geometric parameters as shown in Fig. 1.3.11.

The working length of torsion springs is chosen to lie in the range $l/d = 25 \sim 35$. The internal diameter of the splines D_s at the ends of the torsion element is chosen so that the ratio $D_s/d = 1.2 \sim 1.5$. With these values, the endurance will approach tens of millions of load cycles. To reduce the structural length of torsion elements, they are used in multiples (Fig. 1.3.11, d). Such a

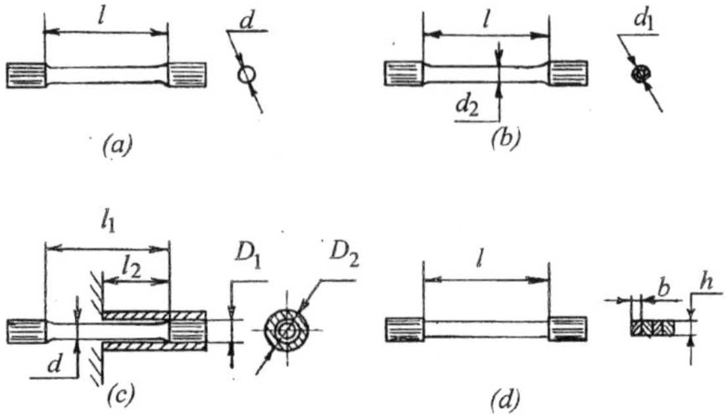

Fig. 1.3.11

layout leads to increased damping in the storage device which is not wanted in resonant drives.

Rubber, rubber-metallic, and elastometric (cellular polyurethane or foam) elastic elements have not been used in resonant drives due to their high energy dissipation (up to 30 ~ 50%).

1.3.5. Energy dissipation.

Energy dissipation when the potential energy store is operating is due to two reasons:

- *Internal friction* in the materials of the elastic component due to its deformation. The nature of internal friction and the way that friction forces depend on the parameters that characterise the deformation of the elastic elements is not at all clear [85]. However, it has been experimentally established that the absorbtion capacity of most materials does not depend on the deformation velocity.

- *Structural damping* due to friction in kinematic pairs of accumulators by the incorporation of certain types of bushes, splines and riveted connections, etc. [107]. Although such assemblies are called stationary, nevertheless, small relative displacements of the contacting surfaces occur where the friction forces produce work. Numerous experiments show that in practice the dissipation properties of these assemblies do not depend on the velocity of loading.

In general, elastic elements for energy storage are elastic-dissipative. They may be described by dynamic characteristics in which separate terms are used for elastic and dissipative loads $P_e(x)$ and $P_d(\dot{x})$

$$P(x,\dot{x}) = P_e(x) + P_d(\dot{x}) \tag{1.3.22}$$

Such separation is conventional, but not physically possible.

The elastic term in (1.3.22) can be replaced by static load characteristics of the form (1.3.10) to (1.3.12), (1.3.15) and (1.3.19) to (1.3.21). The basic dissipative force in the store consists of the viscous resistance force

$$P_{d1}(\dot{x}) = b\dot{x}\,, \tag{1.3.23}$$

and also the dry friction force (Coulomb friction)

$$P_{d2}(\dot{x}) = N_1 \operatorname{sgn} \dot{x}\,, \tag{1.3.24}$$

where b is the friction coefficient and N_1 is a Coulomb friction constant.

To quantify the intensity of the dissipative energy losses it is assumed that the

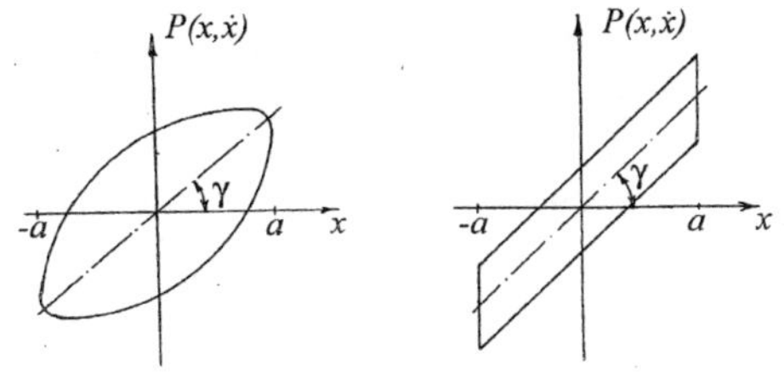

(a) (b)

Fig. 1.3.12

deformation of the dissipative component changes harmonically with amplitude a according to the expressions

$$x = a\cos\omega t,\quad \dot{x} = -a\sin\omega t. \qquad (1.3.25)$$

At the same time, force-displacement diagram for loading and unloading of an elastic element shows hysterisis loops (Fig. 1.3.12). At identical values of deformation, the curve of increasing strain is somewhat higher than that for decreasing strain. The area enclosed by the hysterisis loop represents the energy Ψ dissipated in the elastic element during one cycle of deformation with period $T = 2\pi/\omega$:

$$\Psi = \oint P(x,\dot{x})dx = \oint P_d(\dot{x})\dot{x}dt. \qquad (1.3.26)$$

Fig. 1.3.12, *a* shows the hysterisis loop for a flexible element with dissipative force (1.3.23) and elastic force $P_e(x) = Cx = x\tan\gamma$. The energy loss takes the value

$$\Psi_1 = \pi a^2\omega b. \qquad (1.3.27)$$

Fig. 1.3.12, *b* shows the hysterisis loop for a storage device with a similar elastic force as for the case of dry friction (1.3.24). For this the energy loss will be

$$\Psi_2 = 4aN_1. \qquad (1.3.28)$$

The dissipative properties of materials and structural damping of flexible elements can be conveniently characterised by a *dissipation coefficient*, equal to the ratio between the energy Ψ lost per cycle and the maximum value of the potential energy L. For the linear elastic characteristic (1.3.10) we find $L = 0.5Ca^2$ and the dissipation coefficient is then

$$\psi = 2\Psi/(Ca^2). \qquad (1.3.29)$$

For flexible elements with viscous behaviour (1.3.23) or (1.3.27) and dry friction (1.3.24) or (1.3.28) we find correspondingly

$$\psi_1 = 2\pi\omega b/C,\quad \psi_2 = 8N_1/(Ca). \qquad (1.3.30)$$

Dissipation in flexible elements can also be characterised by *logarithmic decrement* δ related to the dissipation coefficient (1.3.29) by the expression $\psi = 2\delta$, where δ characterises the rate of damping for the natural vibration of the storage device [107]. The simplicity of finding δ experimentally from the amplitude of the relative deformation of the flexible element supports the preferred use of the logarithmic decrement.

Thus, when describing analytically the elastic-dissipative behaviour of the mechanical accumulator, the forces can be written in the form

$$P(x,\dot{x}) = P_e(x) + b\dot{x} + N_1 \operatorname{sgn} \dot{x} . \quad (1.3.31)$$

When searching for a periodic solution of the form (1.3.25), the initial dynamic characteristic (1.3.31) can be replaced by the equivalent visco-elastic model

$$P(x,\dot{x}) \approx Cx + b_0\dot{x} . \quad (1.3.32)$$

The coefficient b_0 in the equivalent damping is chosen so that the energy losses in the initial and equivalent systems are identical. For accumulator devices with viscous behaviour (1.3.23), (1.3.27) or (1.3.30) and with dry friction (1.3.24), (1.3.28) or (1.3.30) we find correspondingly

$$b_{01} = \psi C/(2\pi\omega), \quad b_{02} = 4N/(\pi a\omega) . \quad (1.3.33)$$

The operation efficiency of a mechanical potential energy accumulator is high when ψ and δ are small. For construction purposes it is recommended to choose a high quality spring made from silicon-nickel-vanadium steel. At points where the flexible elements are attached to the components, high contact pressures occur which ensure that the dissipation coefficient is reduced. The importance of satisfying the last requirement is confirmed by the experimental significance of term C_k in (1.3.16), which depends on the form of attachment at the ends of a spiral spring.

1.3.6. Locking of operating components at positioning points. The principal factors in a resonant robotic system are the positioning accuracy, operation at high speed, and the reliability of operation is largely determined by the arrangements for locking the operating element of the robot at the positioning points. In resonant drives, mainly, positioning systems using catches are found (either retractable or rigidly fixed). In these *locking devices* are used to ensure that the elements of a resonant drive are stopped and held in the defined position as self-braking is

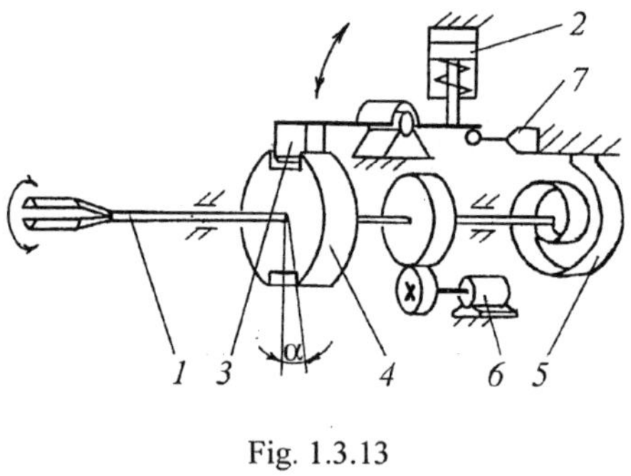

Fig. 1.3.13

applied in the direction of displacement of the restrained element. The stops can be actuated either compulsorily or automatically [73]. Fig. 1.3.13 shows the arm *1* of a robot which can be locked every 180° of rotation about its axis by the action of cylinder *2* which causes latch *3* to enter the indent in element *4*. A wide variety of locking arrangements used in resonant robots: pawls, cams, and ratchets, etc [10, 71, 73].

A study of the methods for locking mechanisms with discontinuous action shows that the positioning error of the drive depends on the accuracy of

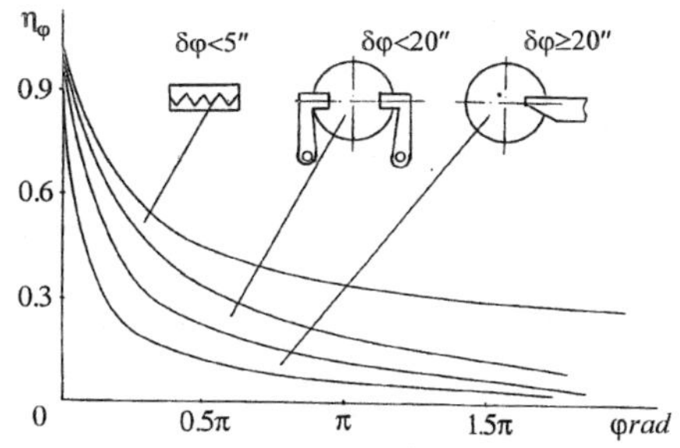

Fig. 1.3.14

adjustment of the latch, the rigidity of the elements and the stability of the spring force on the latch. Thus, the greater the positioning accuracy δ_φ then the higher is the locking-time loss coefficient $\eta_\varphi = t_\varphi / t_f$ where t_φ is the time for vibration to cease after the operating element has been stopped, after which the amplitude of vibration does not exceed the required positioning accuracy δ_φ; t_f is the motion cycle time. Fig. 1.3.14 shows how the locking-time loss coefficient η_φ depends on the rotation angle φ of the operating element for various locking arrangements [12].

For the high-speed cyclic resonant drive shown in Fig. 1.3.13, the maximum deformation of spring *5* occurs at the positioning point and as a result, the locking arrangement is heavily loaded. Considerable friction force acting between the contacting surfaces of the latch *3* and component *4* makes it necessary to use a power cylinder *2*. To reduce energy consumption on releasing the latch and to relieve the latch drive it is necessary to choose the optimum angle of slope α for the detent walls and the latch *3*. A similar effect can be obtained by inputting a variable polarity voltage to the actuator *6* at the instant when the drive is to be released *(Patent USSR 1366388, IPC B25J9/00)*. As a result the friction force between the latch *3* and the walls of the detent in component *4* will be reduced to zero, which facilitates the exit of the latch from the detent. At the instant when the latch is released, a signal from sensor *7* causes the required voltage polarity to be applied to the actuator. The variable polarity can be generated by a capacitor in the control circuit of actuator *6*.

Electromagnetic interactive forces are also used to hold or boost operating components in resonant drives. Fig. 1.3.15 shows the arm of a resonant robot *(Patent USSR 568346, IPC B25J9/00)* in which the locking arrangement is provided by the interaction between a pair of permanent magnets *1* and the electromagnets *2*. Simple control of the electromagnet polarity, to produce not only a force to attract but also to repel the permanent magnet *1* makes it possible to dispense with actuator *3* from the drive. Then, when electromagnet 2 repulses

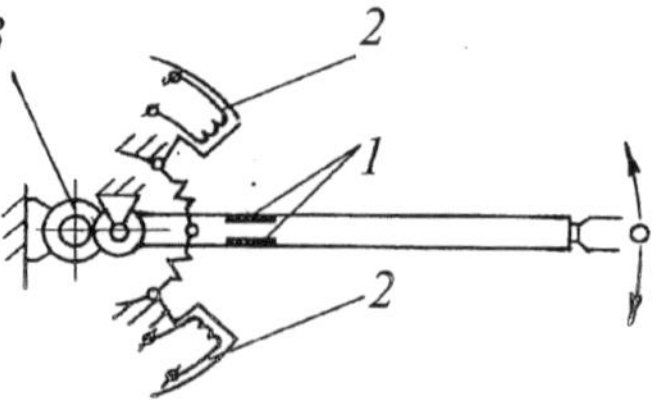

Fig. 1.3.15

magnet *1*, the energy dissipated during motion is supplied to the drive.

The force F generated by the electromagnet is proportional to the square of the current i and depends nonlinearly on the size of the gap between the core and the armature [82]

$$F = \mu_0 S i^2 \Phi(x). \tag{1.3.34}$$

Here μ_0 is the magnetic permeability, S is the area of the armature, $\Phi(x)$ is some nonlinear function (for a uniform magnetic field $\Phi(x) = (\Delta + x)^{-2}$, where Δ and x are found from Fig. 1.3.16). The dynamic characteristic of an electromagnetic locking device $F(u(t), x(t))$ is determined as the system

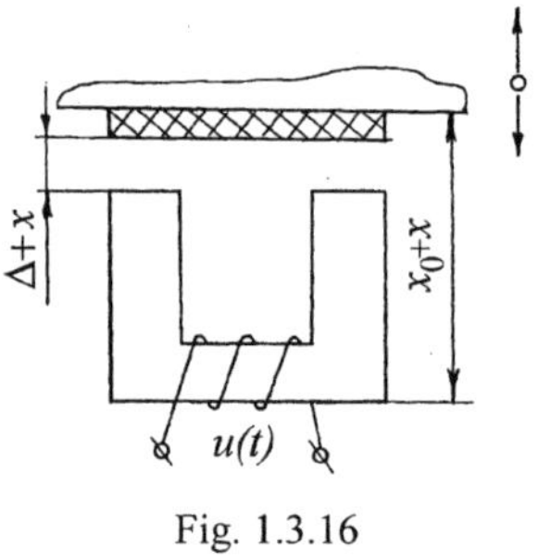

Fig. 1.3.16

consisting of equation (1.3.34) and the equation of an electric circuit

$$\frac{d}{dt}[L(x)i] = u(t) - Ri,$$

where R is the active resistance of the electromagnet and $u(t)$ is the voltage in it.

For secure holding and positioning of moving components in resonant drives electromagnetic disc brakes are widely used.

Fig. 1.3.17 shows a single disc brake with electromagnetic control, locking spring and double friction pairs *(German Patent 2832523)*. The spring closing force generates a maximum braking torque, for which the brake is designed, and the electromagnetic force releases the drive. The friction disc *2* fixed on the output shaft *12* is placed in the body *11*, between the supporting stationary disc *1* and armature *13* of the electromagnet 6 which has coil *5* and core *7*. The aluminium armature *13* contains core *3* and is attracted to the electromagnet when braking is released. The force of spring *8* attracting the core to the braking disc *2* is thus

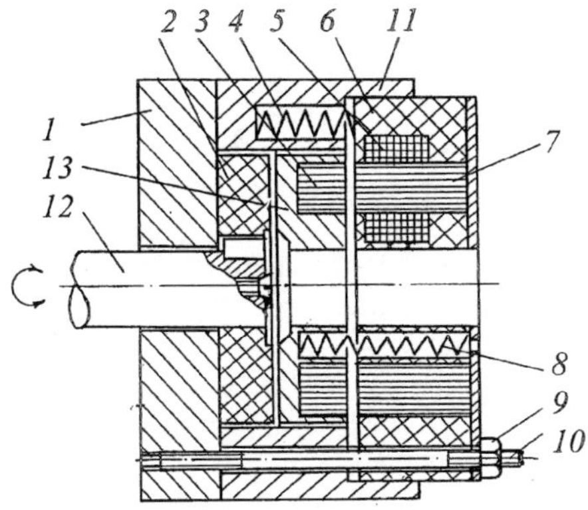

Fig. 1.3.17

overcome. The body and electromagnet are attached to the bearing plate by the bolts *10*. The braking torque between the core and the friction disc is applied to the body through axial flutes on the periphery of the core that mate with flanges on the body. The body is made from material with high damping, for example, reinforced glass-fibre resin. Between the electromagnet and the face of the body is a gap, which is selected by the pre-loading of the nuts *9*. This is used to compensate for wear of the friction surfaces and to maintain the uniform gap between the core and armature of the electromagnet. Spring *4* forces the

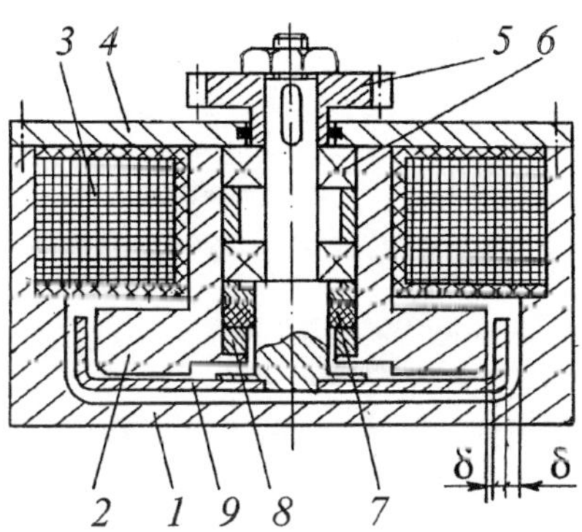

Fig. 1.3.18

electromagnet away from the body.

Recently, there has been a trend to use electromagnetic braking in robots in which the source of braking is an electromagnetic field acting across either dry powder or fluid filling - *powder braking and braking with ferromagnetic fluids* [82]. The effect of such brakes is based on the action of dry or oil-suspended ferromagnetic powders that increase in viscosity in a magnetic field and adhere strongly to the surfaces of the magnet system. Under relative motion of the working surfaces of the brake a resisting force is developed due to the friction between the magnetised particles of the powder. The majority of the motion is experienced by the particles in the middle layers; there is no motion of the particles of the adjacent surfaces to which they adhere and consequently there is no wear on the working surfaces not participating in the friction. When a magnetic field disappears, the resisting friction of the powder or suspension falls strongly and the working components of the brake are no longer connected with each other. Powder brakes have the drawback that their endurance is inferior (200 - 500 hours) due to deterioration and degradation of the powder [40].

Fig. 1.3.18 shows a vertical type of electromagnetic braking system with dry powder filler arranged below [40]. It consists of the body *1*, armature *2*, and cover *4* manufactured from steel. The thin-walled low-inertia steel cylindrical rotor *9* is attached to the output shaft, mounted in the bearings *6*. The excitation windings *3* are arranged around the armature. The cavity between the body and the armature, in which the rotor is arranged, is filled with ferromagnetic powder. The bearing is protected from the penetration of the powder by a reinforced oil-impregnated silicon ring *7* and a stationary annular magnet *8*. The brake is connected to the driven component of the robot drive through the gear *5*. Under the action of a magnetic field, generated by the control winding, the powder within the gap aggregates and becomes compacted. When the rotor surfaces move relative to the walls of the body resistance is developed due to the friction between the magnetised powder particles. The specific effort of motion, which causes a braking torque, depends on the magnetic induction in the operating gap, the composition of the powder and the design parameters for the brake. The output braking torque is virtually proportional to the control current and does not depend on the slipping speed. The residual torque when the control current is absent is not more than 0.5% of the nominal torque.

During the design stage of resonant robotic systems the choice of locking device design is a multi-criteria problem. It is necessary to take into account the peculiarities of the mechanical accumulator: the positioning accuracy requirements, the operating speed and energy consumption during locking.

§1.4 Synthesis of spring accumulators

In resonant drives the type of potential energy accumulator determines to a considerable extent the dynamic character of the drive and the motion of the operating component, and also the type of actuator control and the choice of locking arrangement. Potential energy accumulators consist of two functional parts - the elastic elements and the transmission mechanism, and as a result can be classified as mechanisms with elastic elements [63]. In such mechanisms, the elastic elements are not part of the kinematic chain, but are supplementary to it, linking the separate components of the mechanism. Within the workspace the assumed elastic elements do not impose any limit on the motion of the mechanism and do not alter its degree of freedom. As a result they confer on the mechanism the conditions of a typical vibrating system. The characteristics of basic types of elastic elements have been explained earlier in §1.3.

We shall continue to consider transmission mechanisms.

Transmission mechanisms in the accumulator serve to transmit the forces between the elastic elements and the driven components in direct and reverse directions. The degree of complexity of the transmission mechanism depends on the required motion characteristic for the operating component of the drive.

Before synthesising an accumulator it is necessary to choose the form in which energy will be stored. A comparison can be made of the efficiency of operation of accumulators in terms of energy, overall bulk and process factors: relative energy capacity (J/kg), relative power capacity (W/kg), durability (number of operating cycles) and the coefficient of energy losses (characterising the internal losses in the accumulator).

Accumulator energy indicators

Accumulator type	Energy capacity, kJ/kg	Power capacity, W/kg	Endurance (number of cycles) for 75% discharge
Inertia (flywheel)	$10-6\times10^3$	$>10^4$	$>10^5$
Potential energy - gaseous - plastic - steel spring	 28 3.2 0.32	 $>10^4$ ≈ 80 $>10^4$	 $>10^7$ $10^3-5\times10^3$ $>10^7$
Electro-chemical	60 - 800	80 - 150	300 - 1000

From the table [50] it can be seen that inertia accumulators are the best from the energy capacity standpoint, however their high technological complexity prevents them from being widely applied in practice [46, 74, 80]. Gas accumulators, which have the highest energy capacity amongst potential energy accumulators, require relatively simple control of charging and discharging, which gives them a considerable advantage. However, considerable adiabatic heat losses (up to 40% of the energy) reduce their efficiency. For simplicity and reliability, as well as low internal losses, spring accumulators are widely used.

The following stages in the design of spring accumulators can be identified: structural synthesis, load synthesis, precision and dynamic syntheses. Furthermore, at any stage the synthesis can be defined as an optimisation problem to take into account a few particular quality criteria of the accumulator. Notice that the load synthesis stage for a spring accumulator differs from the load synthesis of a mechanism without flexible links.

By *structural synthesis* of a spring accumulator we will mean the design of the structural scheme of its transmission mechanism according to the given structural conditions. The conditions of motion of the driven component of the accumulator (the degrees of freedom and the type and form of motion) represent structural conditions. The choice of the number of components, the number, class and order of arrangement of the kinematic pairs of the transmission mechanism are the essence of the problem of structural synthesis of a spring accumulator.

Plane linkages, gears and rack-and-pinion mechanisms are used to transmit loads from the elastic elements of the accumulator to the transmission drive. Cam mechanisms have not been used in mechanical accumulators. Note that recently, lever mechanisms have been preferred over cam mechanisms in cyclic machines [111]. Such a choice can be explained by the simpler manufacture of lower kinematic pairs* by comparison with the higher, and by the fact that they provide smooth motion of the operating components. Below, we shall use generally accepted conventions to represent the initial components (Fig. 1.4.1) and kinematic pairs. In Fig. 1.4.2 R and S are respectively the lower rotational and

Fig. 1.4.1

sliding pairs of the fifth class; G is a higher pair of the fourth class supporting rolling motion with sliding of one element of the pair relative to the other (gear meshing) [11]. In our case, the initial component is one of the moving components of the transmission; in particular it can be the drive operating component.

According to the classification used in [11], plane mechanisms with lower kinematic pairs of the fifth class and higher pairs of the fourth class (gear meshing) are plane mechanisms of the third group. The structural formula of PL Chebishev for them takes the form

$$W = 3n - 2p_5 - p_4, \tag{1.4.1}$$

where W is the number of degrees of freedom of the mechanism, n is the number of moving components, p_4 and p_5 are the number of kinematic pairs of the fourth and fifth classes respectively. The number of its generalised coordinates defines the number of degrees of freedom of the driven component of the accumulator. In the majority of cases it is completely satisfactory to use a single-motion transmission mechanism with one degree of freedom. Double motion mechanisms are mainly constructed based on planetary gear systems.

When synthesising structural schemes for accumulators it is necessary to obtain statically determinate systems, which satisfy the assembly technical

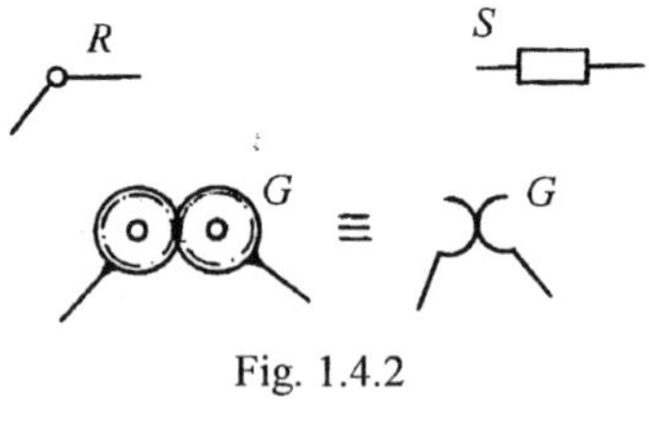

Fig. 1.4.2

* The lower kinematic pair is the connection between a pair of bodies when the relative motion is characterised by two surfaces sliding over one another.

requirements and are independent of the component deformations.

At the *load synthesis stage* for a mechanism with flexible links the problem is solved by choosing the form of static characteristic of the accumulator with constant parameters for its kinematic arrangement and also by the manner of placing the flexible links.

The form of the three static characteristics $L(\varphi)$, the potential energy, $M(\varphi)$, the load characteristic, $j(\varphi)$, the stiffness, for the accumulator predetermine the resonant motion of the operating components of the drive, the actuator control and the form of locking arrangement. The characteristics are related by the expressions

$$M(\varphi) = dL / d\varphi, \ \ j(\varphi) = dM / d\varphi = d^2 L / d\varphi^2 , \qquad (1.4.2)$$

where φ is the displacement of the transmission component to which the accumulator is linked.

Fig. 1.4.3 shows several forms of the static characteristics for accumulators. Characteristics of types *(a)* to *(f)* are for storage devices for stepped drive motion. Locking of the operating component occurs when $M = 0$. The value of the potential energy at this condition is a maximum for types *(a)* to *(d)* and differs little from the maximum for types *(e)* and *(f)*. The segments when $L = const$ and $M = 0$ in curves *(c)* and *(d)* represent the possibility that the operating component can move with constant velocity. In several cases it is necessary to provide a locked position which is locally stable *(e)* and *(f)*. In the remaining cases *(a)* to *(d)* the driven component is locked in the position of unstable equilibrium.

Cyclic motion is achieved by accumulators with characteristics of forms *(g)* to *(o)*. Similarly to the curves *(b)* to *(d)*, accumulators of types *(j)* to *(l)* allow moving the driven component with constant velocity. Locking the driven component in positions where the curve of $L(\varphi)$ has inflexion points, *(h)* and *(k)*, makes it possible to unload the locking device and to reduce the energy expenditure used to control energy storage. Troughs in the curves for $L(\varphi)$, (i) and (l) show that it is possible to obtain cyclic systems in which operating component locking occurs in a stable equilibrium state and the locking device is relieved from the effect of the spring forces. Three-position cyclic systems have characteristics of type *(n)*. The optimal scheme of accumulator with regard to a high-speed criterion has type *(m)* characteristics.

At the force synthesis stage the desired static accumulator characteristics can be obtained for the selected structural scheme of the transmission mechanism by varying relative component sizes and different ways of connecting springs (superposing flexible links).

In the force synthesis of accumulators it is convenient to use the position function and the three transfer functions known from the theory of cyclic mechanisms [101]. In the same way, the characteristics for energy storage with a linear spring can be incorporated.

From the condition of equal work on the possible displacements of the accumulator we can write

$$M(\varphi) = \Pi'_p(\varphi) Cx \qquad (1.4.3)$$

where φ is the displacement of the driven component of the accumulator, C and x are the stiffness and deformation of the spring, Π'_p is the first transfer function of the energy store, defined by the position function $x = \Pi_p(\varphi)$ through its derivative

$$\Pi'_p(\varphi) = d\Pi_p / d\varphi . \quad (1.4.4)$$

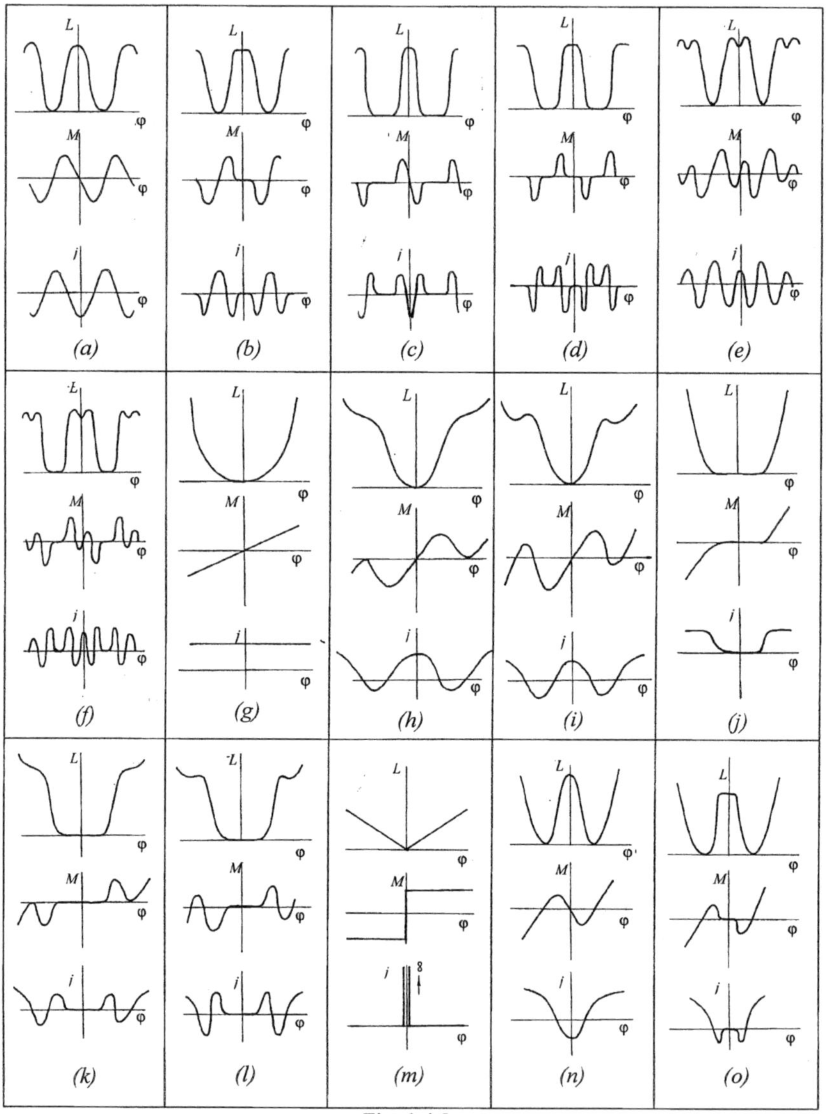

Fig. 1.4.3

Using the second transfer function of the mechanisms [111]

$$\Pi''_p = d^2\Pi_p / d\varphi^2, \tag{1.4.5}$$

we write the characteristic for accumulator stiffness

$$j(\varphi) = dM / d\varphi = C(\Pi'_p)^2 + C\Pi''_p x. \tag{1.4.6}$$

For a number of transmission mechanisms in which the position functions are known, expressions (1.4.3) to (1.4.6) simplify the force synthesis of accumulators.

At the *precision synthesis stage* the conclusion about the adequacy of the transmission mechanism is based on the desired precision of the static characteristics. Corrections must be made at the earlier load synthesis stage when the transmission mechanism does not meet the precision criterion. If it is not possible to implement corrections, different structural schemes for the accumulator should be considered

At the *dynamic synthesis stage* it is necessary to match the dynamic parameters of the accumulator: the mass and stiffness of the components, and also the stiffness, energy requirements and attachment conditions of the elastic elements. The solution of this problem is based on the analysis of the dynamic model of the drive as a whole. The required velocity and energy values of the resonant drive are used as primary dynamic conditions for the synthesis. In view of the complexity of constructing an analytic solution, in the majority of cases numerical modelling is carried out.

Optimisation synthesis consists of finding the constant parameters of the transmission mechanism to ensure that the optimisation criterion extreme is met. For this, account must be taken of the limits imposed on the acceptable values of the parameters sought. Constant values substituted for the parameters, determine the kinematic and dynamic scheme of the mechanism or its structure. Depending on the stage at which the optimisation synthesis is being carried out, the optimisation criteria and parameters sought will change. The most commonly used methods for mechanism optimisation are the random search method, the sampling search method, the LP-optimisation method and others [81].

The conditions that the accumulator must satisfy can not always be formulated mathematically in the form of optimisation criteria. Frequently this is concerned with the structural requirements of the transmission mechanism. The most important criteria for the comparison of different types of accumulators for potential energy are: the operation time for a single displacement; overall dimensions, simplicity; the degree to which the motion of the driven component corresponds to the initial requirement; the kinematic characteristics of the transmission mechanism; the energy requirement of the accumulator control system; the presence of "hard or soft" impacts during operation; the possibility of installing the accumulator in position sufficiently close to the operating component of the drive; the peak value of the reaction transmitted to the frame; the reliability and endurance of the accumulator.

We shall concentrate on the first two stages of synthesis (structural and force) for one-degree-of-freedom energy accumulator with linear springs. We shall not make a comprehensive synthesis of the structural arrangement of the transmission mechanism since this is well carried out in the literature [9, 11, 28, 56, 58, 59, 70, 73, 89, 94, 95, 98, 101, 109, 112, 114, etc.]. We shall turn our attention to the types of mechanism, which give the static characteristics shown in Fig. 1.4.3, and

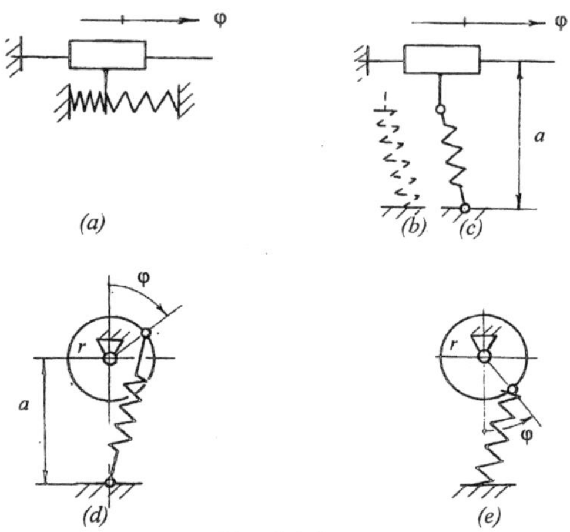

Fig. 1.4.4

also we shall examine several mechanisms, which have already been widely used in practice.

1.4.1 Elements of synthesis for two- and three-component accumulators[2].

The simplest potential energy store can be obtained by using elastic connections between the initial components (Fig. 1.4.1) and the frame. As an example let us consider the two-component energy store capable of forming cyclic (Fig. 1.4.4, *(a)* to *(c)*) or stepped machine motion (Fig. 1.4.4, *(d)* and *(e)*). Fig. 1.4.4 and other figures of the current chapter show kinematic arrangements for accumulators without actually showing how the kinematic pairs and components can be practically embodied. For example, in Figs. 1.4.4 *(d)* and *(e)* the driven component of the accumulator could be a gear meshing with one of the transmission drive components. This can also be achieved when the driven component of the accumulator is also the operating component of the resonant drive. Accumulators with linear elastic characteristics (Fig. 1.4.3, *(g)*, Fig. 1.4.4, *(a)*) are widely used in practice [4, 7, 20]:

$$L = C\varphi^2/2,\ M = C\varphi,\ j = C. \qquad (1.4.7)$$

In the case of a rotating initial component, the elastic link with the frame consists of a spiral spring.

A three-position cyclic machine (Fig. 1.4.3, *(n)*) can be constructed using a spiral compression spring and a linear initial component (Fig. 1.4.4, *(b)*): $l > a$

$$\begin{aligned} L &= 0.5C[l-(a^2+\varphi^2)^{1/2}]^2, \\ M &= C\varphi[1-l(a^2+\varphi^2)^{1/2}], \qquad (1.4.8)\\ j &= C[1-l(a^2+\varphi^2)^{1/2}]-Cl\varphi^2(a^2+\varphi^2)^{-1/2}, \end{aligned}$$

[2] By two- and three-component accumulators we mean spring accumulators with two, three or more components in the transmission mechanism. The frame is considered as one of the accumulator components.

Here and below, the following nomenclature will be used: l is the initial length of the spring, a is the specified size of the accumulator, shown in the appropriate figure, r is the radius of the circle to which the spring end is attached.

Locking of the driven component of a resonant drive with the spring as shown in Fig. 1.4.4, *(b)* in the middle position is obtained by the relieved lock.

Depending on the choice of the relative sizes of l and a when using a sliding initial component and tension spring (Fig. 1.4.4, *(c)*) it is possible to achieve characteristics as shown in Fig. 1.4.3, *(g)* or *(j)*:

$$\begin{aligned} L &= 0.5C[(a^2+\varphi^2)^{1/2}-l]^2, \\ M &= C\varphi[1-l(a^2+\varphi^2)^{1/2}], \\ j &= C[1-l(a^2+\varphi^2)^{1/2}]-Cl\varphi^2(a^2+\varphi^2)^{-1/2}. \end{aligned} \tag{1.4.9}$$

The most widespread scheme of an energy store for a stepped system is shown in Fig 1.4.4, *(d)* [21]. When $l<a$ the accumulator characteristics take the form (Fig. 1.4.3, *(a)*)

$$\begin{aligned} L &= 0.5C[r^2+a^2+2ar\cos\varphi)^{1/2}-l]^2, \\ M &= -Car[1-l(r^2+a^2+2ar\cos\varphi)^{-1/2}]\sin\varphi, \\ j &= -Car\{[1-l(r^2+a^2+2ar\cos\varphi)^{-1/2}\cos\varphi- \\ &\quad -lar\sin^2\varphi(r^2+a^2+2ar\cos\varphi)^{-3/2}\}. \end{aligned} \tag{1.4.10}$$

At the locking point, when $\varphi=0$ the accumulator stiffness is negative and takes the value

$$j|_{\varphi=0}=-Car[1-l(r+a)^{-1}]. \tag{1.4.11}$$

The use of a compression spring and a rotating component when $l>a$ makes it possible to construct a stepped energy store (Fig. 1.4.3, *(a)*, Fig. 1.4.4, *(e)*) *(Patents USSR 1357219, IPC B25J9/00, and 1516689, IPC F16H33/02)* with characteristics of the form

$$\begin{aligned} L &= 0.5C[l-(r^2+a^2-2ar\cos\varphi)^{1/2}]^2, \\ M &= Car[1-l(r^2+a^2-2ar\cos\varphi)^{-1/2}]\sin\varphi, \\ j &= Car\{[1-l(r^2+a^2-2ar\cos\varphi)^{-1/2}]\cos\varphi+ \\ &\quad +lar\sin^2\varphi(r^2+a^2-2ar\cos\varphi)^{-3/2}\}. \end{aligned} \tag{1.4.12}$$

The stiffness of the accumulator in the locked position has the value

$$j|_{\varphi=0}=Car[1+l(r-a)^{-1}]. \tag{1.4.13}$$

Using the simplest two-component energy store (Fig. 1.4.4) it is possible to achieve cyclic and stepped motion, and to unload the lock and construct a three-positional cyclic drive. However, the use of a three-component energy store gives more possibilities of realising different regimes and relieving locking devices. Alongside the structural formula (1.4.1), for a three-component mechanism the following condition must be satisfied:

$$p_5+p_4=3. \tag{1.4.14}$$

Bearing in mind that one of the components of a three-component mechanism is the frame, we can write a formula for the number of varieties of dyads N_p as a function of the number of variable kinematic pairs p

$$N_p = p^3 - pC_p^2, \tag{1.4.15}$$

where C_p^2 is the number of combinations of the number p of kinematic pairs in twos [68]. Fig. 1.4.5 shows the possible construction of dyads for $p=3$ (two pairs of the fifth class and one gear pair). The dotted line surrounds Assur's groups [11]. From formulas (1.4.1) and (1.4.14) it follows that three-component one-degree-of-freedom mechanism ($W=1$) is generated when a single gear pair is used ($p_4=1$) with two pairs of the fifth class ($p_5=2$).

Fig. 1.4.6 shows several forms of three-component energy stores.

The simplest is the double-crank store *RGR*, consisting of two connected gears (Fig. 1.4.6, *(a)*) on the base of which is constructed a resonant drive according to *Patent USSR 1453108 IPC F16H33/02*, described in section 1.5. The characteristics of this store are shown in Fig. 1.4.3, *a* and can be described in the form

$$\begin{gathered} L = 0.5C(a-l+2r\cos\varphi)^2, \\ M = -2Cr(a-l)\sin\varphi - 2Cr^2\sin 2\varphi, \\ j = -2Cr(a-l)\cos\varphi - 4Cr^2\cos 2\varphi. \end{gathered} \tag{1.4.16}$$

The accumulator stiffness at the locking position will be

$$j|_{\varphi=0} = -2Cr(a-l+2r). \tag{1.4.17}$$

Replacing the rotational pair in the mechanism of Fig. 1.4.6, *a* by a sliding pair produces the mechanism *RGS* shown in Fig. 1.4.6, *b*, which ensures that the lock is unloaded during cyclic work:

$$\begin{gathered} L = 0.5C\{r[(\varphi+\sin\varphi)^2 + (1+\cos\varphi)^2]^{1/2} - l\}, \\ M = Cr\varphi(1+\cos\varphi)\{r - l[(\varphi+\sin\varphi)^2 + (1+\cos\varphi)^2]^{-1/2}\}, \\ j = Cr(1+\cos\varphi - \varphi\sin\varphi)\{r - l[(\varphi+\sin\varphi)^2 + (1+\cos\varphi)^2]^{-1/2} + \\ + Crl\varphi^2(1+\cos\varphi)^2[(\varphi+\sin\varphi)^2 + (1+\cos\varphi)^2]^{-3/2}. \end{gathered} \tag{1.4.18}$$

Similar static characteristics (Fig. 1.4.3, *h*) are possessed by the *RGS* mechanism shown in Fig. 1.4.6, *c*. The expressions for the characteristics due to displacement φ_1 can be obtained from (1.4.18) by substituting $\varphi = \varphi_1 r^{-1}$.

Various accumulator arrangements can be obtained by using planetary mechanisms. Segments of the path on which the driven component moves at constant velocity (Fig. 1.4.3, *c*) can be obtained from stepped motion, by using a planetary three-component transmission mechanism, as shown in Fig. 1.4.6, *d* [9, 59, 71, 73]. Here, energy storage must satisfy the expression $k = r_2 r_1^{-1} = 3, 5, 7, \ldots$ The characteristics shown in Fig. 1.4.3, *e*, *f* occur when $k = r_2 r_1^{-1} = 4, 6, 8, \ldots$ Such a store is shown in Fig. 1.4.6, *e*. Here it is possible to obtain characteristics of the form as shown in Fig. 1.4.3, *f* (by using elastic links at points *D* and *B*).

The mechanisms drawn in Fig. 1.4.4, *d* and *e* and in Fig. 1.4.6, *a* have the static characteristics shown in Fig. 1.4.3, *a*, which, although they belong to the same type, are different from each other. Frequently, it is required that the driving torque grows rapidly at exit from the position of unstable equilibrium. The acceleration intensity of the driven component depends on the value of the accumulator stiffness in the locked position. As can be seen from formulas (1.4.11), (1.4.13) and (1.4.17) the accumulator stiffness is greatest for that of Fig. 1.4.4, *e* and least for that of Fig. 1.4.4, *d*.

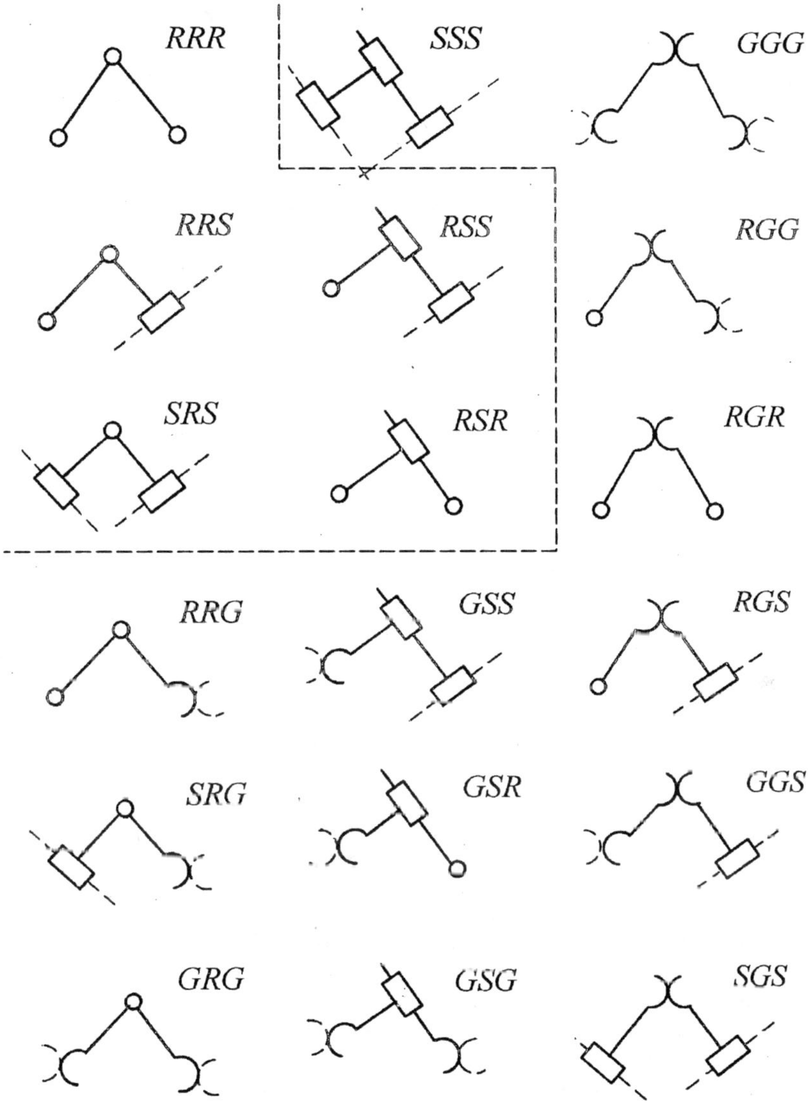

Fig. 1.4.5

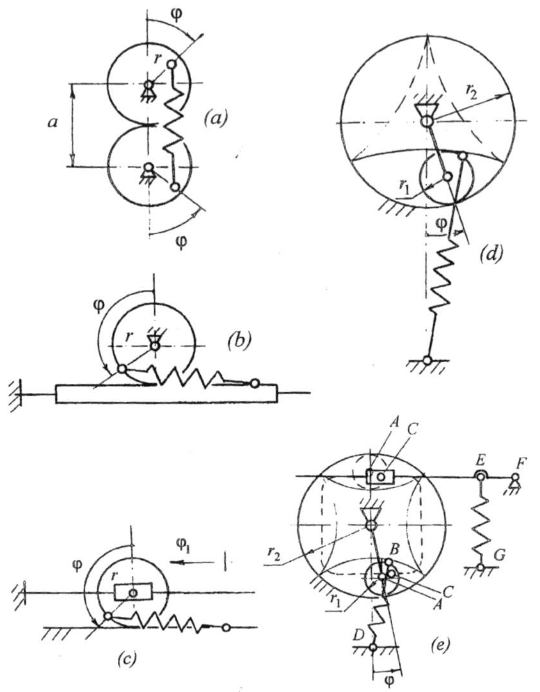

Fig. 1.4.6

Analysis of the practical operation of two- and three-component energy stores shows that in most cases they can ensure load relieving of robot cyclic drives. Their simplicity and reliability together with their wide operating conditions makes them preferable choices during the structural synthesis of potential energy accumulators.

1.4.2. Multi-component spring accumulators.

The need to examine multiple components, which play a role in transmission mechanisms of energy storage devices, is dictated by the wide spectrum of requirements imposed on accumulator mechanisms. For example, complex motion, with stops and reversals, can not be achieved by simple two- or three-component energy stores.

Following [9, 58], by convention all multi-component mechanisms can be separated by function into guidance or transmission applications and also by the degree of accuracy and precision with which they characterise the function as approximate or precise mechanisms. In *guidance* mechanisms, a point on the component undergoing complex motion, reproduces in advance the given curve or part of it on a plane attached to the frame. *Transmission* mechanisms provide the required functional link between the input and output components which is described by the transfer function of the mechanism. If the components are inter-related then the link is called a position function. If there is a relationship between the velocities of the input and output components then the connection is called the transfer ratio (e.g. gear ratio). If the given curve (functional link) coincides with a curve (functional link) reproduced by the mechanism on the whole plane, then such a mechanism is called a *precise* guidance (transfer) mechanism. If the reproduced mechanism curve (the functional link) is sufficiently close to the given curve over a limited interval, differing little from it more or less at all the remaining points of the plane, then such a mechanism is called an *approximate* guidance (transfer) mechanism. The rare examples of precise mechanisms contain a small number of components. For this reason, in practice, mechanisms are approximated. The majority of the structural arrangements quoted below are related to approximate mechanisms.

The most representative and well investigated is the four-bar linkage mechanism [9, 56, 59, 76, 114]. To correspond with the examples of reasoning laid out above we shall look at a four-bar linkage mechanism of the third group. From structural formula (1.4.1) when $W=1$, we find the condition

$$2p_5 + p_4 = 8\,. \tag{1.4.19}$$

In addition, using the fact that the number of kinematic pairs of the fourth and fifth classes can be between four and six [9], another condition can be written

$$p_5 + p_4 = q,\quad \text{where } q = 4,\ 5,\ 6. \tag{1.4.20}$$

Using (1.4.19) and (1.4.20) we can present in tabular form the possible combination of kinematic pairs

q	4	5	6
p_4	0	2	4
p_5	4	3	2

Taking into account that the number of such combinations is quite large we shall limit ourselves below to consider only those types of transmission mechanisms that have been widely used in practice.

The sinusoidal mechanism (Fig. 1.4.7) with a spring, is used in many vibration

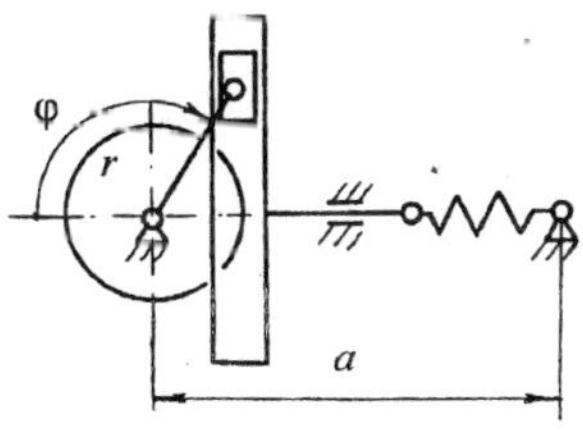

Fig. 1.4.7

machines [9, 10, 71, 73]. The static characteristics (Fig. 1.4.3, *a*) for the store based on sinusoidal mechanism almost coincide with the characteristics (1.4.16) for a double-crank store (Fig. 1.4.6, *a*)

$$L = 0.5C(a - l + r\cos\varphi)^2,$$
$$M = -Cr(a - l)\sin\varphi - 0.5Cr^2\sin 2\varphi, \qquad (1.4.21)$$
$$j = -Cr(a - l)\cos\varphi - Cr^2\cos 2\varphi.$$

Using an ellipsograph (Fig. 1.4.8) makes it possible to relieve the lock by

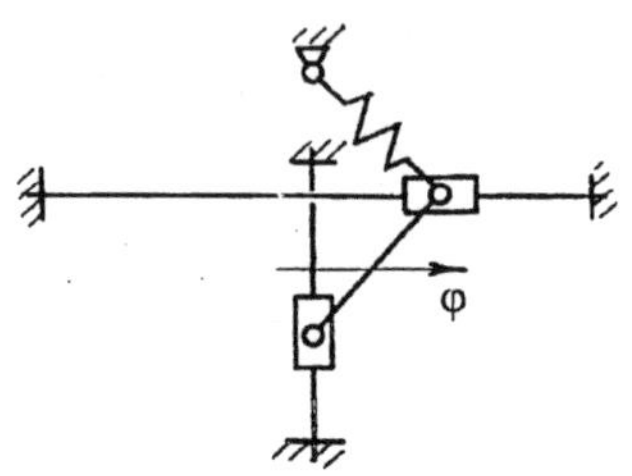

Fig. 1.4.8

loading the intermediate lever. The characteristics for this mechanism similar to (1.4.9) are drawn in Fig. 1.4.4, *c*.

By using a slider-crank mechanism it is possible to obtain various types of

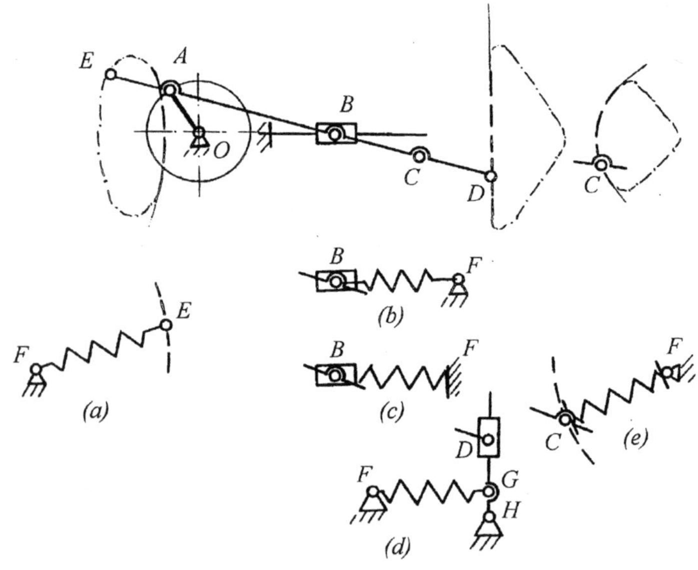

Fig. 1.4.9

motion for the driven component. By constraining one end of the spring to a lever moving through some trajectory with either linear or rotational motion, and the other end to the frame, it is possible to obtain the characteristics shown in Fig. 1.4.3 *a* to *c*. The choice of the spring attachment points is decided by graphical or analytic methods [9, 28, 59, 76, 94, 114]. Static characteristics when $a > l$,

$AB = h$, $OA = r$, $OF = a$, $\varphi = \pi - B\hat{O}A$ and the spring is attached at point B take the form

$$\begin{gathered} L = 0.5C[a - l + r\cos\varphi - (h^2 - r^2\sin^2\varphi)^{1/2}]^2, \\ M = -Cr(a-l)\sin\varphi - Cr^2\sin 2\varphi + Cr\sin\varphi(h^2 - r^2\sin^2\varphi)^{1/2} + \\ +0.5Cr^2\sin 2\varphi(a - l + r\cos\varphi)(h^2 - r^2\sin^2\varphi)^{-1/2}, \\ j = -Cr(a-l)\cos\varphi - 2Cr^2\cos 2\varphi + Cr(h^2 - r^2\sin^2\varphi)^{-1/2}\times \\ \times(h^2 - 2r^2\sin^2\varphi)\cos\varphi + 0.5Cr^2\sin 2\varphi(h^2 - r^2\sin^2\varphi)^{-1/2}\times \\ \times\{2(a - l + r\cos\varphi)\cot 2\varphi + 0.5r^2(a - l + r\cos\varphi)\times \\ \times(h^2 - r^2\sin^2\varphi)^{-1}\sin 2\varphi - r\sin\varphi\}. \end{gathered} \tag{1.4.22}$$

When a compression spring is used (Fig. 1.4.9, *c*) the characteristic for $\varphi = B\hat{O}A$ takes the form

$$\begin{gathered} L = 0.5C[l - a + r\cos\varphi + (h^2 + r^2\sin^2\varphi)^{1/2}]^2, \\ M = -Cr(l-a)\sin\varphi - Cr^2\sin 2\varphi - Cr\sin\varphi(h^2 - r^2\sin^2\varphi)^{1/2} - \\ -0.5Cr^2\sin 2\varphi(l - a + r\cos\varphi)(h^2 - r^2\sin^2\varphi)^{-1/2}, \\ j = -Cr(l-a)\cos\varphi - 2Cr^2\cos 2\varphi - Cr(h^2 - r^2\sin^2\varphi)^{-1/2}\times \\ \times(h^2 - 2r^2\sin^2\varphi)\cos\varphi + 0.5Cr^2\sin 2\varphi(h^2 - r^2\sin^2\varphi)^{-1/2}\times \\ \times\{2(l - a + r\cos\varphi)\cot 2\varphi + 0.5r^2(l - a + r\cos\varphi)\times \\ \times(h^2 - r^2\sin^2\varphi)^{-1}\sin 2\varphi - r\sin\varphi\}. \end{gathered} \tag{1.4.23}$$

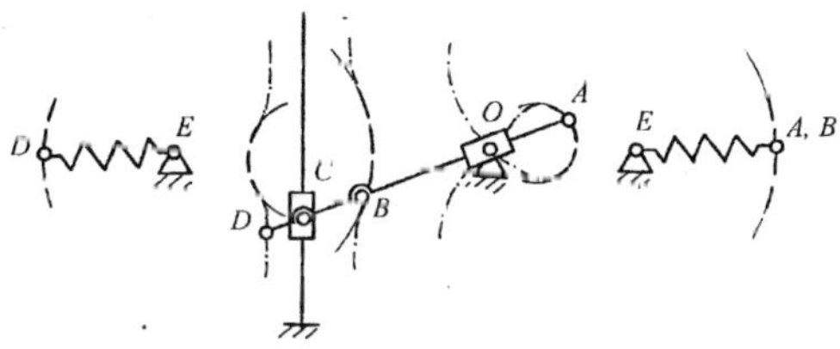

Fig. 1.4.10

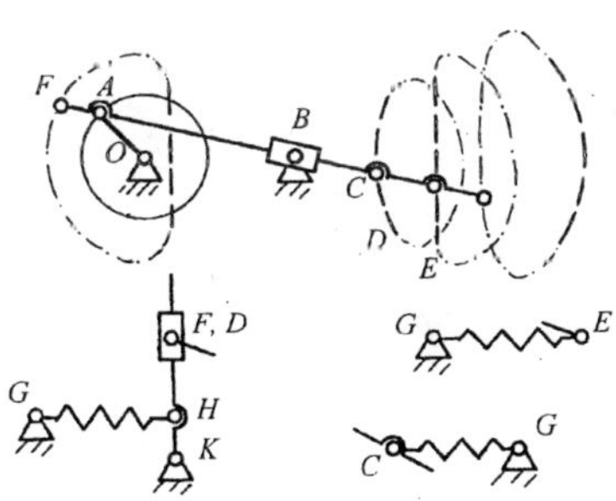

Fig. 1.4.11

The conchoidal mechanism (Fig. 1.4.10) with a spring attached to the points *A*,

B or *D* that follow approximately through circular arcs at some portion of trajectory, provides constant velocity of the driven component over these parts.

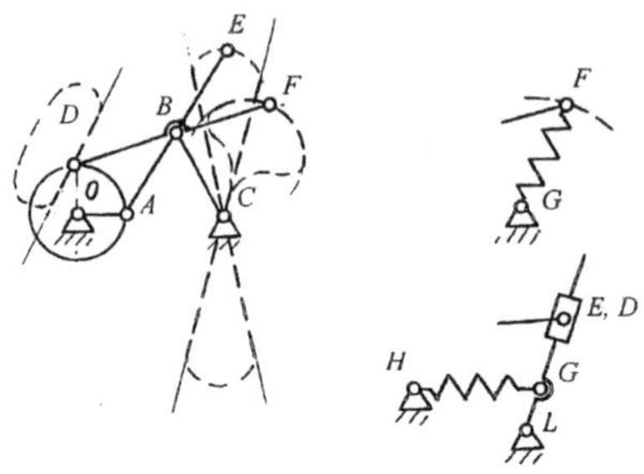

Fig. 1.4 12

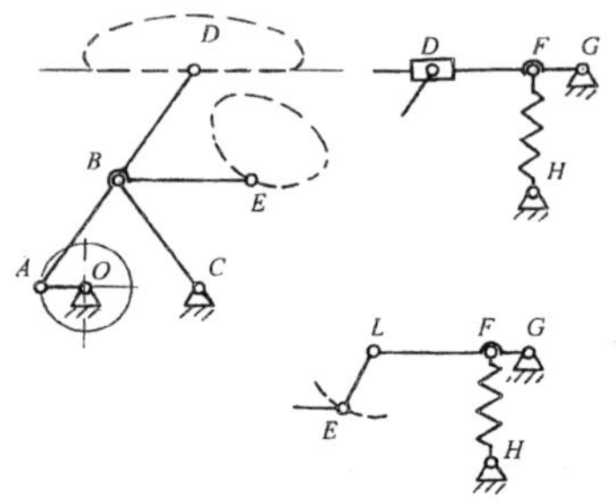

Fig. 1.4.13

The characteristics for the mechanism are drawn in Fig. 1.4.3, *j* and *o* [59].

The crank-and-rocker mechanism (Fig. 1.4.11) has similar properties to the

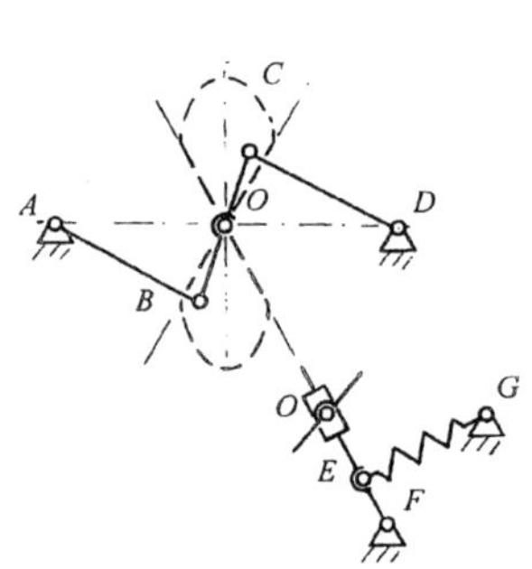

Fig. 1.4.14

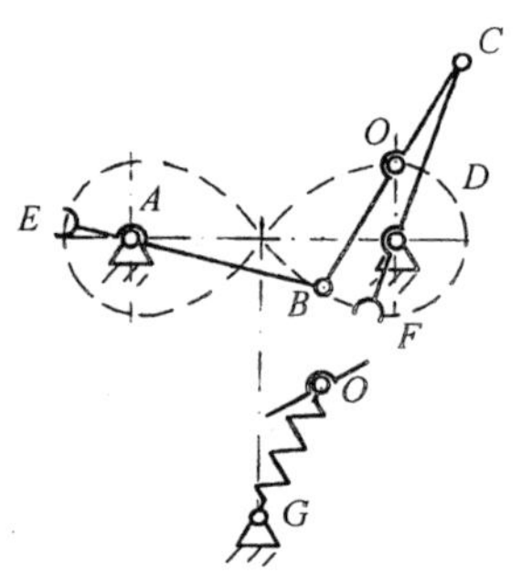

Fig. 1.4.15

slider-crank mechanism (Fig. 1.4.9) [28, 59, 71].

In practice the commonest is the four-bar linkage mechanism, consisting of

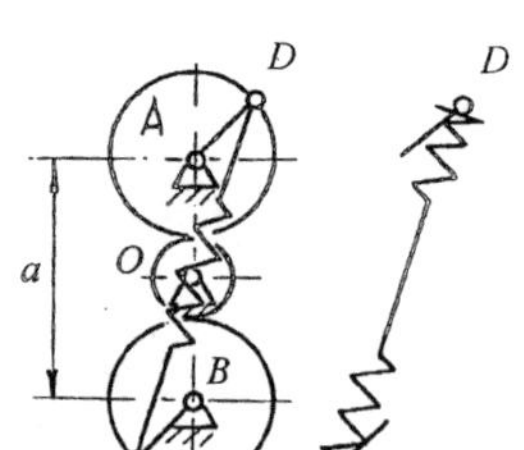

Fig. 1.4.16

single rotating pairs. Using this mechanism as an approximate guide either for linear or rotation arc motion, it is possible to construct an energy store with the characteristics shown in Fig. 1.4.3 *a* to *c*, *h* and *j*. Several such practical stores based on four-bar linkage mechanisms are shown in Fig. 1.4.12 to 1.4.15. The kinematic design for an approximate guidance (transfer) mechanism is carried out by well-known methods [9, 28, 59, 71, 77, 94, 114].

An energy store constructed from two kinematic pairs of the fourth class and three of the fifth class is shown in Fig 1.4.16 *(Patent USSR 1516689, IPC F16H33/02)*. Its characteristics correspond to those drawn in Fig. 1.4.3, *a*

a) $\varphi = \pi - O\hat{A}D = \pi - O\hat{B}C$

$$\begin{aligned} &L = 0.5C[(a^2+4r^2+4ar\cos\varphi)^{1/2} - l]^2,\\ &M = -2Car[1 - l(a^2+4r^2+4ar\cos)^{-1/2}]\sin\varphi,\\ &j = -2Car[1 - l(a^2+4r^2+4ar\cos)^{-1/2}]\cos\varphi +\\ &+4Cla^2r^2\sin^2\varphi(a^2+4r^2+4ar\cos\varphi)^{-3/2}\sin^2\varphi. \end{aligned} \tag{1.4.24}$$

b) $\varphi = O\hat{A}D = O\hat{B}C$

$$\begin{aligned} &L = 0.5C[-(a^2+4r^2-4ar\cos\varphi)^{1/2} + l]^2,\\ &M = 2Car[1 - l(a^2+4r^2-4ar\cos)^{-1/2}]\sin\varphi,\\ &j = 2Car[1 - l(a^2+4r^2-4ar\cos)^{-1/2}]\cos\varphi +\\ &+4Cla^2r^2\sin^2\varphi(a^2+4r^2-4ar\cos\varphi)^{-3/2}\sin^2\varphi. \end{aligned} \tag{1.4.25}$$

The planetary gear mechanism shown in Fig. 1.4.17 has two springs of stiffnesses C_1 and C_2. Its characteristics are similar to those in Fig 1.4.3, *a*:

$$\begin{aligned} &L = 0.5C_1(a - l_1 - r\cos\varphi)^2 + 0.5C_2(a - l_2 + r\sin\varphi)^2,\\ &M = C_1r(a-l_1)\sin\varphi + C_2r(a-l_2)\cos\varphi + 0.5(C_2 - C_1)r^2\sin 2\varphi,\\ &j = C_1r(a-l_1)\cos\varphi - C_2r(a-l_2)\sin\varphi + (C_2 - C_1)r^2\cos 2\varphi. \end{aligned} \tag{1.4.26}$$

It is worth noting that when $a = l_2$ then the second spring has no pre-load in it, and the elastic characteristic of the store is similar to the mechanical

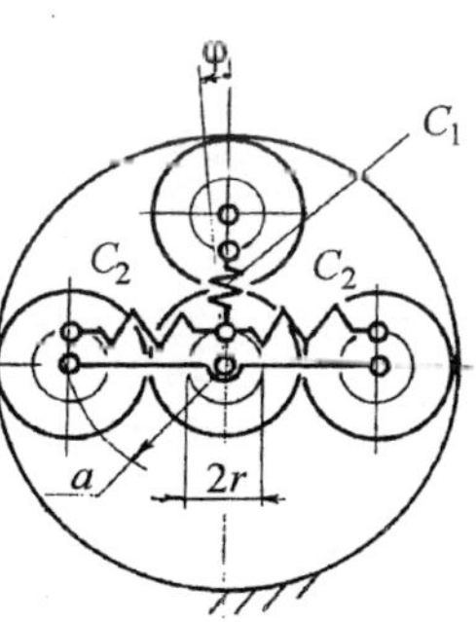

Fig. 1.4.17

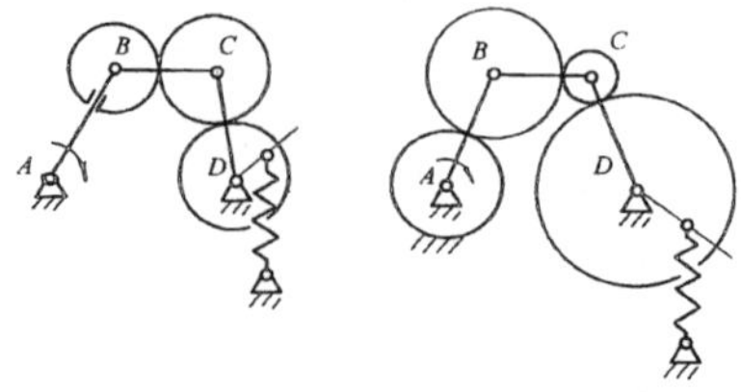

Fig. 1.4.18

characteristics of a two-phase explicit-pole synchronous electrical motor [37]

$$M = C_1 r(a - l_1)\sin\varphi + 0.5(C_2 - C_1)r^2 \sin 2\varphi \,. \qquad (1.4.27)$$

The parameters for a mechanical device and an electrical machine, ignoring the actual dimensions, are related to each other as follows

$$\varphi = \theta,\ r = e,\ a - l_1 = E_d,\ C_1 = (\omega_0 x_d)^{-1},\ C_2 = (\omega_0 x_q)^{-1}, \qquad (1.4.28)$$

where θ is the working angle of lag of the rotor from the stator field, e is the initial voltage in the stator windings, E_d is the excitation voltage, x_d and x_q are the inductive resistances, ω_0 is the frequency of the supply voltage.

In practice all the characteristics drawn in Fig.1.4.3 can be obtained using gear-linkage mechanisms [95] for example, as shown in Fig. 1.4.18. However, the synthesis of gear-linkage energy stores is coupled with highly complex calculations and requires separate in-depth study.

In §3.2 during the course of solving an optimisation problem the elastic characteristic (3.2.24) is obtained for an energy store that provides a drive

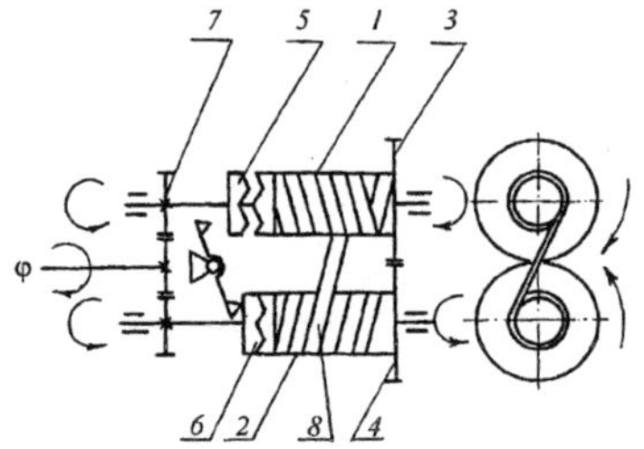

Fig.1.4.19

operating at high speed with low energy consumption on control. Relay characteristics of the form (3.2.24) as in Fig. 1.4.3, *m* can be obtained by using a cam transmission mechanism. However high contact stresses prevent the construction of a durable and reliable accumulator. It is more effective to construct an energy store based on a spring mechanism (Fig. 1.4.19) such as *Patent USSR 180045 IPC F16H33/02* and *Patent USSR 203411 IPC F16H33/06*, which both ensure constant torque. The accumulator consists of a master drum *1* and slave drum *2*, connected to each other at one end by gears *3* and *4*, and at the other end by the clutches *5* and *6* attached to the transmission *7*. A helical plane spring *8* is used as an elastic element, one of whose ends is attached to and coiled around the master drum *1*, while the other end is rigidly attached to the slave drum *2* and wound in the opposite direction. During acceleration of the operating component through clutch *5*, a constant driving torque M_0 is transmitted from the master drum *1* to the transmission. As the operating component passes through the mid-point $0.5\varphi_T$, clutch *5* opens and clutch *6* closes, and the energy store loads the transmission with a torque M_0 from the slave drum *2*.

The foregoing study of the structural arrangement of stores represents a very small portion of the many varying types of mechanical accumulator which can not be considered here but can be obtained using the Roberts/Chebishev transformation or a combination of non-Assur groups [58, 59].

It must be mentioned that the feasibility of implementing multi-component accumulators without a deep study of force analysis of transmission mechanisms remains open. It is necessary to specify the tolerance on the maximum pressure angle (or the minimum tolerance on the gear angle) arising from the conditions under which the mechanism will be used.

Structural synthesis of energy accumulators gives the designer several schemes of device that satisfy some or other initial requirements. Because of the high frame reactions, preference is given to mechanisms that cause symmetrical forces on the frame, for example as in Fig. 1.4.6, *a* and 1.4.16.

§1.5 Design of resonant robotic systems

The conditions for using resonant robotic systems can vary considerably. During the design of resonant systems for particular operating conditions two approaches can be applied. The first is based on choosing special mechanical energy accumulators that ensure that the motion of the operating component agrees closely to that programmed. As mentioned earlier, the accumulator for this design is very complex. The other design approach consists of using a simple accumulator with additional control devices (clutch, selector or brake, etc.) in the kinematic chain of the accumulator - operating component. In this case the control system of a resonant robot becomes more complex. For example, displacement of the operating component can be achieved on the parts of the motion with constant velocity by using the accumulators shown in Figs. 1.4.6, *d* and *e*, 1.4.9, 1.4.10 and 1.4.11. This can also be achieved using simple types of accumulator (Figs. 1.4.4, 1.4.6, *a* to *c* with additional clutches and brakes to disconnect the accumulator from the operating component of the robot over the required portions of the trajectory.

To ensure the operational integrity of robotic systems, it is particularly important to specify the tuning mechanisms and to construct the mechanical part of the drive properly. Above all, this concerns the arrangement of a neutral position for the elastic system of the accumulator and the method for changing (in some cases, automatically) the positioning point of the operating component. For example these requirements apply to the resonant drive of automatic guided vehicles. Their stopping points can be not only at processing stations and locations to recharge an electric accumulator, but also at any position along the motion path.

By using several spring accumulators in one drive the processing capabilities of resonant robotic systems can be enhanced. By switching these sequentially or simultaneously the distance and intensity of acceleration and braking can be varied.

It is possible to increase the dynamic quality of resonant robot operation by using the effect of dynamic interaction between simultaneously operating resonant drives with several degrees of freedom. It also appears effective in some cases to kinematically link the components of the resonant drive with the moving components of the basic process equipment. For example, a built-in resonant robot feeder, kinematically connected to the carriage of a machine tool, recharges the energy dissipated by motion.

The complex character of operation of resonant robotic systems must be taken into account at the machinery design stage. We shall consider several types of design solutions that make it possible to increase the dynamic and operational quality, reliability and working capacity of resonant robots.

1.5.1 Methods for the mechanical tuning and control of resonant drives.

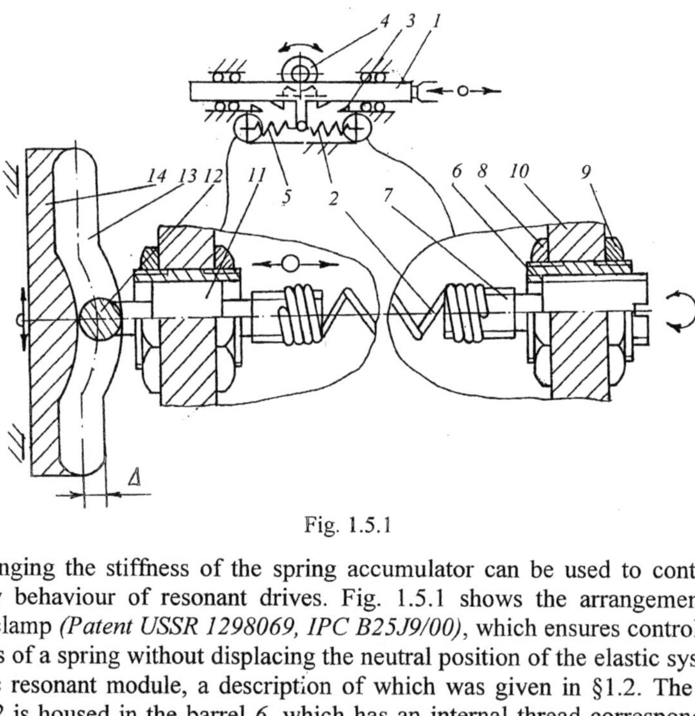

Fig. 1.5.1

Changing the stiffness of the spring accumulator can be used to control the velocity behaviour of resonant drives. Fig. 1.5.1 shows the arrangement of a spring clamp *(Patent USSR 1298069, IPC B25J9/00)*, which ensures control of the stiffness of a spring without displacing the neutral position of the elastic system of a cyclic resonant module, a description of which was given in §1.2. The spring clamp *2* is housed in the barrel *6*, which has an internal thread corresponding to the thread on the anchor with pitch t the same as the pitch of the spring in the neutral position of member *1*. The barrel 6 is connected to the frame *10* and can move axially within it. The barrel simplifies the initial setting of the anchor (e.g. aligning the pitch of the spring with pitch t of the threaded end fitting *7* with simultaneous neutral setting of the operating component *1*). After the initial position of the barrel *6* has been established, the nuts *8* and *9* are locked, and control of the spring stiffness is accomplished by rotating the end fitting *7*. By matching the thread pitch in the end fitting *7* to the pitch of spring *2* at the neutral position of the operating component *1* it is possible to change the number of working turns and to vary the spring stiffness without altering the neutral position

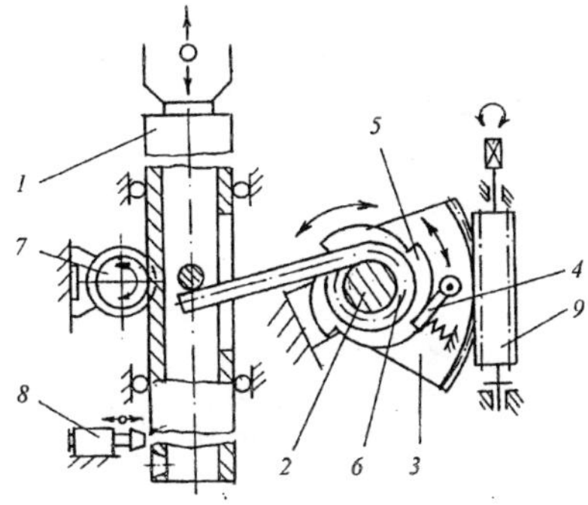

Fig. 1.5.2

of component *1*. Using this clamping method it is unnecessary for the pair of springs *2* and 5, used in one drive, to have identical kinematic parameters.

The arrangement of a cyclic resonant drive (Fig. 1.5.1) can be considerably simplified by choosing the length of the springs *2* and *5* so that in the free state, in any position of the operating component *1*, the forces on it from both springs are non zero *(Patent USSR 1511107, IPC B25J9/00)*. For this to occur, the springs must remain in opposition, and their lengths are chosen from the condition

$$l_c > 0.5(a-y)+x, \quad l_t < 0.5(a+y)-x, \qquad (1.5.1)$$

where l_c and l_t are the lengths of the compression and tension springs in the free condition; a is the distance between the fixed end of springs *2* and *5* from the body *10*; x is the amplitude of displacement of the operating component *1* from the stable equilibrium position; y is the distance of the fixed ends of springs *2* and *5* attached to the operating component *1*. Satisfying condition (1.5.1) ensures symmetrical overall stiffness characteristics for springs *2* and 5, which eases the resonant arrangement of the drive.

Fig. 1.5.2 shows the arrangement for a cyclic resonant lifting module for the operating component *1* with a mechanism for accurately controlling the static equilibrium position *(Patent USSR 1386444, IPC B25J9/08)*. The mechanism consists of a gear sector *3*, with sprung pawl *4* and ratchet wheel *5* attached to it. This is mounted on pivot *2*, and internal spiral spring *6*, one end of which is attached to ratchet *5*, while the other acts on the operating component *1*.The drive is energised by the actuator *7*. Latch 8 locks member *1* in its extreme position. The gear sector meshes with worm wheel *9*.

As the weight carried changes, the static equilibrium of the module alters. Thus, when the weight increases the module is displaced until the increasing elastic force in spring *6* as it is twisted balances the weight of loaded member *1*. To re-establish the static equilibrium position it is necessary to increase the initial

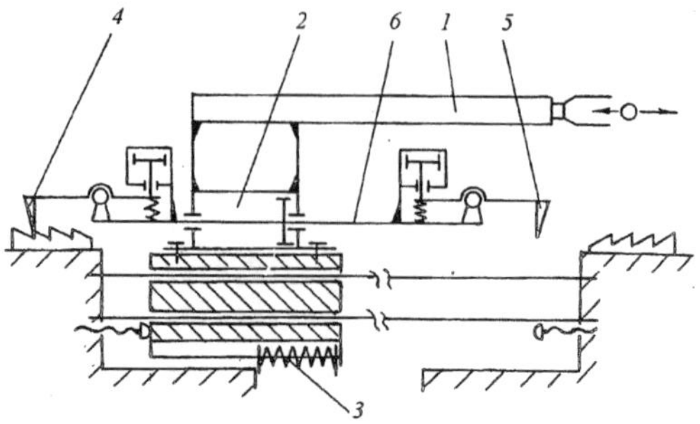

Fig. 1.5.3

preload in spring *6*. Coarse adjustment is supplied by rotating the ratchet wheel *5* with subsequent locking of the pawl *4*. Accurate adjustment is made by rotating the ratchet wheel *5* by means of the worm *9* while pawl *4* remains fixed. The displacement of the pawl is transmitted to the ratchet and it is rotated to the angle required for the corresponding original static equilibrium position. The schematic arrangement shown allows variation of the static load within wide limits and preserves the resonant arrangement of the drive.

In resonant drives, friction losses are not constant during operation. Environmental temperature changes, wear and similar processes change the losses in the drive. A resonant state is characterised by the equilibrium between energy losses during motion and replenishment from the driver. Such balance is in general ensured by means of adaptive control (Chapter 4). Technical solutions are also known which eliminate hunting of the operating component about the locking point. Fig. 1.5.3 shows the arrangement of a cyclic drive with spring accumulator *(Patent USSR 1466934, IPC B25J9/08)* in which the drive cylinder *2* replenishes dissipative losses by means of the additional deformation of spring *3* at the end of motion of member *1*.

The cyclic module operates in the following manner. In the initial condition, operating component *1* is held by the left-hand latch *4* with spring *3* cocked and the left-hand space of cylinder *2* under pressure. Motion towards the right starts when the left-hand latch is released. Simultaneously the pressure in cylinder *2* is relieved. Operating component *1* undergoes acceleration, deceleration and braking. At the end of braking, member *1* is engaged by the right-hand latch and simultaneously pressure is applied to the right-hand space of cylinder 2 which ensures that, due to the fixed shaft *6*, the body of cylinder 2 moves to the right until operating component *1* reaches the required position. As a result, spring *3* is additionally deformed and its energy increases by that of the dissipative losses. To compensate possible friction losses the mating parts of the latches *4* and *5* consist of geared racks. Cylinder *2* is a double-acting hydraulic cylinder with small stroke but capable of producing the high operating loads necessary to provide the additional deformation to cock the spring.

The energy replenishment in the drive can be governed by means of a computer (optimal and adaptive control, see Chapters 3 and 4) or mechanically as

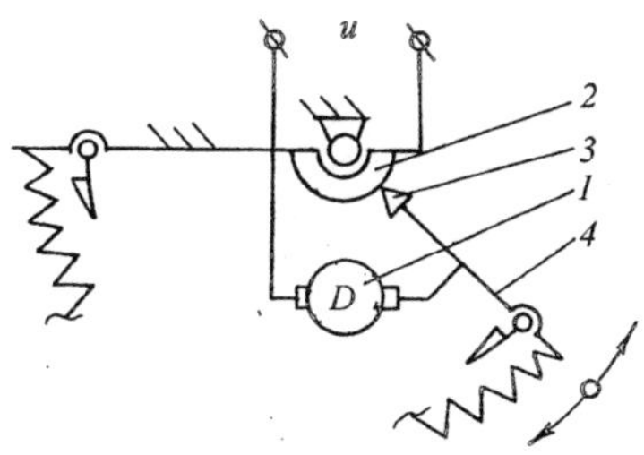

Fig. 1.5.4

shown in Fig. 1.5.4. The voltage u from the controller *(Patent USSR 1151449, IPC B25J9/00)* is supplied to actuator *1* through potentiometer *2*, which has a wiper *3* connected to operating component *4*. The voltage u, which ensures the replenishment of energy to the drive, is predetermined by the profile of potentiometer *2*. Such an arrangement for controlling the voltage can limit the motion of member *4* to ensure no overshooting when the displacement changes.

1.5.2 Cyclic resonant drives with multi-position points.

Fig. 1.5.5 shows a cyclic drive *(Patent USSR 1219335, IPC B25J9/00)* in which the pivoted spring latches *2* and *3* are mounted on operating component *1*, as shown in Fig. 1.4.4 *(b)*, which, together with movable stops *4*, allow a

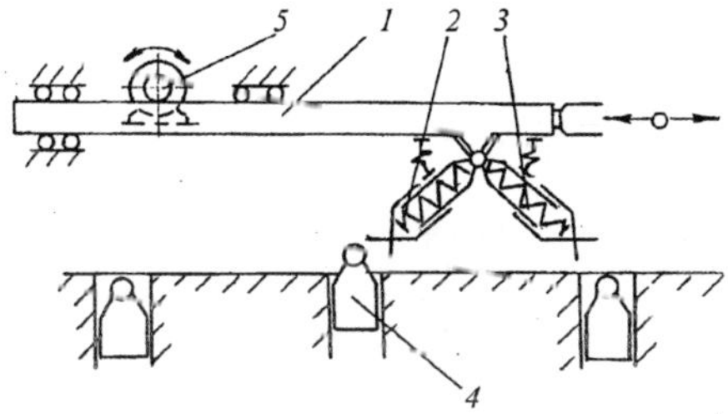

Fig. 1.5.5

considerable increase in the number of positioning points. During operation when movable stop *4* contacts either of springs *2* or *3*, the spring become compressed and rotates about the pivot. The drive is locked in an unstable equilibrium state, which corresponds with a positioning point of operating component *1*. The drive can be initiated from a locked position in any direction by means of the actuator *5*.

A multi-position resonant drive with springs configured according to Fig. 1.4.4 *(a)* is shown in Fig. 1.5.6 *(Patent USSR 1454685, IPC B25J9/00)*. As well as the latches *7* and *8* for the extreme position the drive can take up an intermediate position when stops *4* bearing movable latches *6* are extended. Operating component *1* has reciprocating motion. Motion takes place until latch *6* is driven into the slot *10* of the operating component. The movable stops *4* can be programmed to halt the operating component as required.

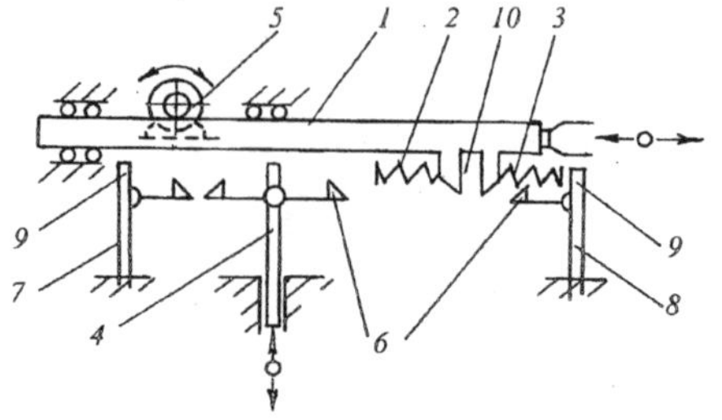

Fig. 1.5.6

Fig. 1.5.7 shows an arrangement for a multi-position rotary spring manipulator module *(Patent USSR 1472250, IPC B25J9/00)*. Fixed gripper *1* can rotate about

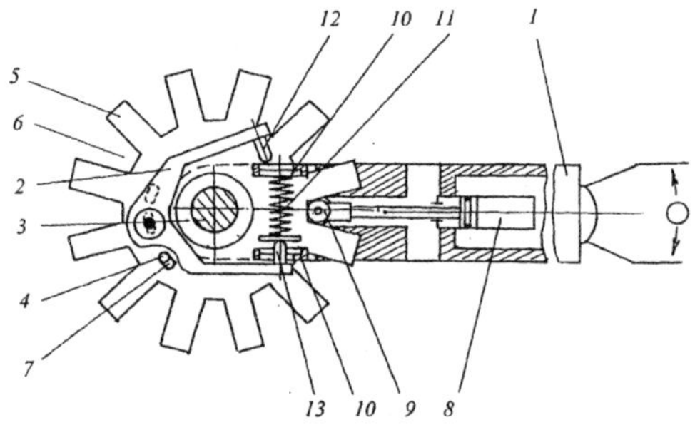

Fig. 1.5.7

axis *3* together with the two brackets *2* on one of which placed the movable latch *4*. Disc *5,* which has equally spaced radial slots *6* on it periphery and is rigidly attached to the axis *3*, also has in it an aperture *7* that engages with latch *4*. Gripper *1* contains a cylinder *8* with a roller *9* on the end of its piston rod, which engages with one of the slots *6* in disc *5*. Compression spring *11* is arranged between the detents *10* in gripper *1* and acts through pads with the ends *12* and *13* of bracket *2*.

The module operates in the following manner. In the initial position (shown in Fig. 1.5.7) the piston of cylinder *8* is extended and roller *9* is inside the slot *6* of disc *5*, latch *4* is engaged in the aperture *7* of bracket *2* and spring *11* is fully compressed. Rotation of gripper *1* under the action of spring *11* begins when roller *9* is retracted from slot *6*. At the instant when contact ceases between end *13* of the bracket *2* and spring *11*, latch *4* is withdrawn from slot *7*. As a result, gripper *1* together with the bracket rotates anti-clockwise due to inertia around axis *3* and disc *5*. As the required position approaches latch *4* enters one of the apertures *7*. Bracket 2 becomes restrained relative to disc *5* while gripper 1 continues to move compressing spring *11*. The kinetic energy of moving gripper *1*is transformed into potential energy of compressed spring *11*. At the same as latch *4* becomes engaged, cylinder 8 moves roller *9* into the nearest slot *6*. As roller *9* moves towards its limit, the spring deforms additionally to replenish the energy expended on dissipation during motion. During the motion cycle, gripper 1 follows three phases: firstly, start up due to the spring action, secondly, rotation due to inertia with weak deceleration and thirdly, braking with energy replenishment. Gripper *1* interacts with the spring during the first and third phases of motion. The gripper moves with reciprocating swinging motion. The number of positioning points is determined by the number of slots *6* in disc *5*.

The use of planetary gears in machine transmissions increases the reliability and functionality, reduces weight and decreases the clearances required, so increasing their productivity and endurance [109]. Planetary gears used in transmissions make it possible to change the intensity of run up and braking in resonant robotic devices. Fig. 1.5.8 shows the drive for a resonant transport

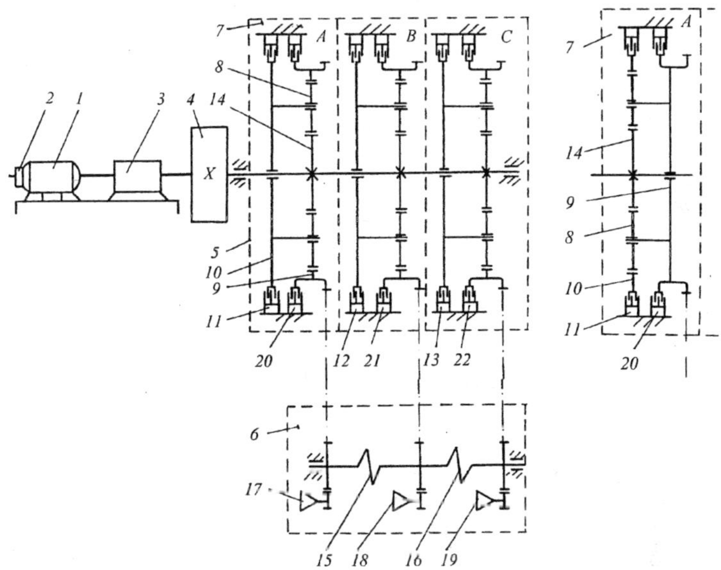

Fig. 1.5.8

vehicle with planetary energy store *(Patent USSR 1453109, IPC F16H33/02)*. The drive consists of actuator *1*, transmission *3*, energy store *6*, which is supplied by a successive combination of elastic elements *15* and *16* which, through mechanical transmission *5*, consisting of identical planetary chains *7*, is attached to the operating component *4*. Each planetary sequence *7* together with the satellites *8* contains three elements *9*, *10*, and *14* consisting of sun gear, spider and central wheel with internal teeth. In each planetary chain, one transmission element is attached to the end of one of the elastic elements of the energy store and of the two remaining transmission elements, one is attached to the controlled brake and the other to the operating component *4*. When the machine stops, braking is transmitted in the sequence *12, 13, 20* to ϕ, *11, 21, 22* to *I* where ϕ and *I* mean that the brake is off or on respectively. Rotation of the operating component is transmitted to element 9 via transmission element *14* and satellites *8*. The kinetic energy of the moving machine is transformed into potential energy of deformation of the elastic elements. At the instant of stopping, as shown by the tachogenerator *2*, all brakes are engaged. Run up of the machine from the charged store occurs by engaging a combination of brakes, which depends on the state of the energy store and the required motion drive. At the instant of acceleration, when the deformation of the elastic elements becomes zero, as shown by the position

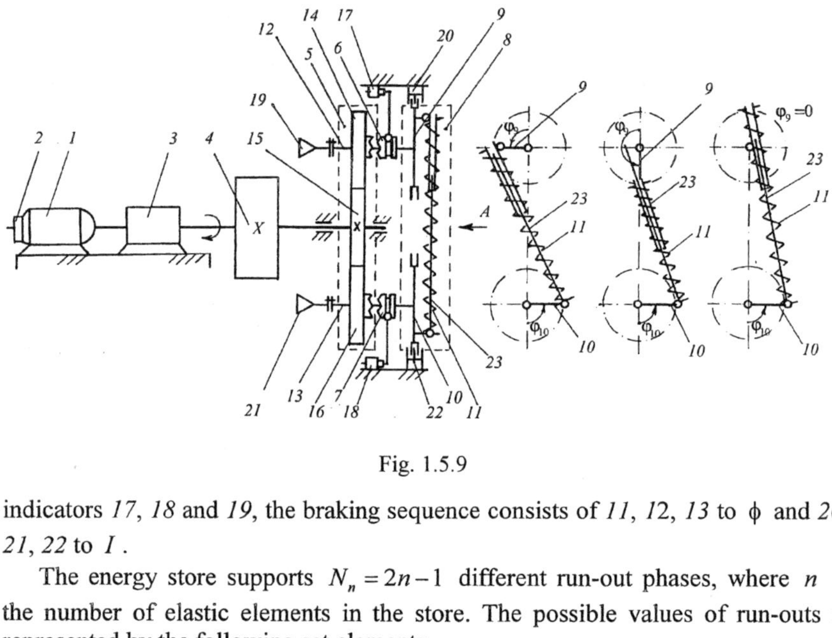

Fig. 1.5.9

indicators *17*, *18* and *19*, the braking sequence consists of *11*, *12*, *13* to ϕ and *20*, *21*, *22* to I .

The energy store supports $N_n = 2n-1$ different run-out phases, where n is the number of elastic elements in the store. The possible values of run-outs is represented by the following set elements

$$\mathfrak{P} = \left\{ln^{-1/2};\ l(n-1)^{-1/2};\ \dots;\ l;\ l(n-1)^{1/2};\ ln^{1/2}\right\},$$

where l is the run-out value when one elastic element is in operation.

1.5.3 Resonant drives with variable positioning

Such drives are used in automatic stackers and guided vehicles.

Fig. 1.5.9 shows a resonant drive for a transport machine *(Patent USSR 1516689, IPC F16H33/02)* with accumulator as shown schematically in Fig. 1.4.4, *d* and *e*.

The drive consists of an actuator *1* with a velocity sensor - tachogenerator *2*, transmission *3*, attached to the operating component *4* and through transmission mechanism *5*, control clutches *6* and *7* with energy store *8*, consisting of cranks *9* and *10* whose shafts *12* and *13* are parallel. Spring *11* is attached to the arms of the cranks. Transmission mechanism *5* contains three gears *14*, *15* and *16*. Gears *14* and *16* are free to rotate about shafts *12* and *13*, but gear *15* is attached to actuator *1* and meshes with gears *14* and *16*. The engagement of clutches *6* and *7* to lock the gears *14* and *16* onto shafts *12* and *13* is carried out by electromagnets *17* and *18*. Shaft *12* and crank *9* are attached to position sensor *19* and is prevented from rotating by control brake *20*. Shaft *13* and crank *10* are connected to position sensor *21* and prevented from rotating by control brake *22*. Tachogenerator *2* and position sensors *19* and *21* together with the control brakes *20* and *22* are connected to the drive control system.

The drive operates in the following manner. Run up of the machine is performed by actuator *1* when accumulator *8* is uncharged. For this, clutches *6* and *7* are released, while brakes *20* and *22* lock cranks *9* and *10*. To provide braking of the machine, actuator *1* is disconnected together with control brakes *20* and *22*, while electromagnets *17* and *18* engage clutches *6* and *7*. Rotation of operating component *4* is transmitted to shafts *12* and *13* and to cranks *9* and *10* through gears *15*, *14* and *16* and clutches *6* and *7*. As they rotate through identical angles, cranks *9* and *10* approach one another and compress spring *11*. The kinetic energy of the machine is transferred into potential energy of deformation in the spring. The machine velocity is strongly reduced. At the instant when the cranks have rotated through and angle $\varphi_9 = \varphi_{10} = 180°$, clutches *6* and *7* are disengaged simultaneously by a signal from position sensors *19* and *21* and brakes *20* and *22* applied. As a result the maximum charged accumulator *8* is securely locked in a position of unstable equilibrium, securely locked, and member *4* continues to rotate with lower velocity. Depending on the velocity of the machine, braking can be implemented by one of cranks *9* or *10*. If machine braking is performed by *10*, then brake *22* is disengaged and clutch *7* engaged. As this takes place, brake *20* is engaged and clutch 6 disengaged. Rotation of member *4* through gears *15* and *16* and clutch *7* is transmitted to shaft *13*. Machine velocity is reduced. The machine is run up as follows. Store *8* is connected to member *4* by engaging clutches *6* and *7* and disengaging brakes *20* and *22*. Loaded spring *11* rotates cranks *9* and *10* which supply a torque to member *4*.

In this way the drive can ensure operation of the machine with positional

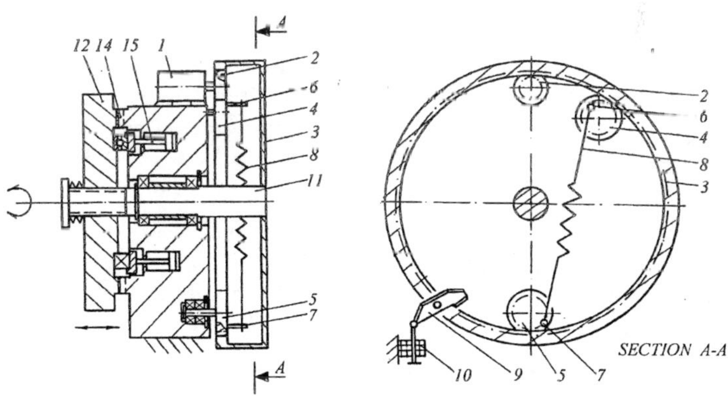

Fig. 1.5.10

braking, and ensures that the machine can run up in forward or reverse directions with up to three levels of velocity with two different rates of velocity change.

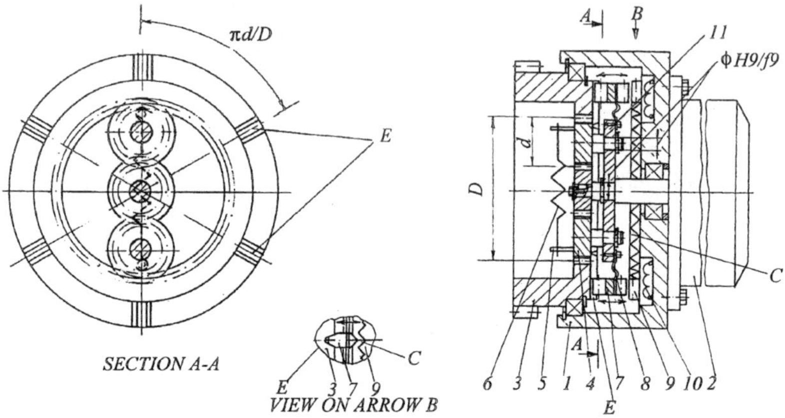

Fig. 1.5.11

1.5.4 Stepped resonance systems. In automated manufacturing many mechanisms (robotic machines, revolving turrets, etc.) operate in stepped modes. By using relieving principles it is possible to increase the speed of operation and reduce energy consumption.

Fig. 1.5.10 shows the drive for a revolving turret *(Patent USSR 1544550, IPC B25J9/00)* with a double rotating crank energy store (Fig. 1.4.16). The drive consists of actuator *1* with driving gear *2*, operating component *3* which is a crown wheel with internal gearing, and an energy store consisting of cranks *4* and *5* with arms *6* and *7* whose ends are connected by tension spring *8*. Operating component *3* can be locked by double lever *9*, driven by electromagnet *10*, and as a result tool disc *12*, which sits on shaft *11*, can also be locked and connected with body *13* by means of the end plate of toothed clutch *14*. Axial motion of disc *12* is provided by hydraulic cylinder *15*.

In the initial stage spring *8* is extended to its maximum and arms *6* and *7* are at the maximum distance apart.

To move disc *12* to the next operating position, hydraulic cylinder *15* moves it away from body 13 by sliding it along the splines on the shaft *11* and as a result, clutch *14* is disengaged and a command is sent to disengage the lock *9* and engage actuator *1*. The drive proceeds from unstable equilibrium. The motion of disc *12* occurs due to the action of torques supplied by the actuator and spring *8*. The transmission ratio is $t = T_4 / T_3 = T_5 / T_3$ where T_j is the number of teeth on the $j-$th element in Fig. 1.5.10. The ratio is chosen so that for a complete rotation of the cranks *4* and *5*, disc *12* rotates through an angle corresponding to its displacement from one position to the next; for example, for an eight-tool turret this angle is 45°.

When the cranks *4* and *5* rotate up to 180° disc *12* accelerates as spring *8* compresses, but when spring *8* is extended, disc *12* decelerates. For a complete revolution of the cranks, the disc can be fixed by lever *9*. Simultaneously, actuator *1* is disengaged while clutch *14* is engaged placing disc *12* accurately onto the body *13*. Under these conditions, spring *8* is extended and the energy store charged.

Using a double-crank energy store as shown in Fig. 1.4.16 it is possible to design structural schemes for the turret in which the body is loaded symmetrically by torques.

With multi-step conditions for a revolving turret, the velocity and torque (Fig. 1.4.3, *a*) are pulsed which has adverse effects on the endurance and reliability of a machine. This deficiency is made good by relieving the drive using a resonant motor-gearbox as shown in Fig. 1.5.11. The gearbox is a double-crank accumulator (Fig. 1.4.16) in which the body of the gearbox is coupled to the rotating actuator *2* and to a toothed disc *9* with locating grooves *C*. The output pinion *3* mates with gears *4* that are interconnected by spring 6 and the central gear is attached to the actuator shaft *2*. Diaphragm *8* with double-sided lock *7* is attached to the spider *11*, which rotates freely on the actuator shaft. Lock *7* interacts with the locating grooves *C* on the toothed disc *9* and the recesses *E*, machined on the body of gear *3*. Lock *7* is controlled by electromagnet *10*. In the initial position of the gearbox, lock *7* is in the recesses *E* of gear *3*. The performance of a single step of motion (rotation of the output gear *3* through an angle $2\pi d/D$ in any direction) occurs by removing lock *7* from recesses *E* and inserting it into grooves *C* using magnet *10* and engaging actuator *2*. As a result the gearbox becomes unstable, with spider *11* braked. For the first half turn, gears *4* are accelerated due to the action of spring *6* and actuator *2*. During the second half-turn of gear *4* the kinetic energy is transferred to potential energy in the spring. At the end of rotation, current is removed from electromagnet *10* and actuator *2* and lock *7* enters into the nearest recess *E* in pinion *3*. Spring *6* is cocked, the gearbox is braked and ready for the next movement.

Multi-step operation in the initial phase of motion is similar to that described previously. When gears *4* rotate for a half turn, electromagnet *10* is disengaged and lock *7* returns into recess *E*. Further rotation occurs from actuator *2* with locked gears *4* and spider *11*. At the end of motion gears 4 are unlocked by electromagnet *10*. After a half-turn of gear *4*, spring *6* is fully cocked. Locking the gearbox is as described earlier. Thus in a motor-gearbox, multi-step motion is not accompanied by torque pulses.

1.5.5 Application of mechanical accumulator combinations in a resonant robotic system. The use of such applications makes it possible to arrange run up and braking processes of different intensity and also to limit motion with different velocity levels *(Patent USSR 323602, IPC F16H33/00)*.

Fig. 1.5.12 shows the scheme for the resonant drive of an automatic guided vehicle *(Patent USSR 1453108, IPC F16H33/02)* with two accumulators: one linear and one as shown in Fig 1.4.6, *a*. When the machine is braked without accurate positioning driver *1* is disconnected and potential energy accumulators *7* and *6* are connected to the operating component either in sequence or simultaneously. Accumulator *6* is engaged using electromagnet *16*, which simultaneously interacts with the clutch *13* and stop *10* on the moving end of elastic element *8*. Accumulator *7* is engaged by electromagnet *27*, which simultaneously acts on clutch *23* with stop *25*. Rotation of operating component *4* is transmitted through mechanical transmission *5* and clutch *13* on the moving end of elastic element *8*, to twist element 8. Rotation of member *4* is transmitted to cranks *18* and *19* through clutch *23*, gears *22* and *21*. The cranks rotate and extend spring *20*. As a result the kinetic energy of the guided vehicle is transferred into potential energy in accumulators *6* and *7* and its velocity is reduced to zero. At the instant when the machine has completely stopped, as shown by tachogenerator *2*, electromagnet *16* is de-energised and clutch *13* is released and stop *10* is locked to the moving end of elastic element *8*. The completely charged accumulator *7* is in an unstable equilibrium state, and automatically disconnected from the operating component and fixed by stop *25*. In this state, the torque at the output from accumulator *7* is zero, and stop *25* is relieved. Run up of the guided vehicle takes place as driver *1* is disengaged and accumulators *6* and 7 are engaged with the driven member. Fully discharged, accumulator *7* is automatically disconnected, and accumulator *6* is disconnected on a signal from position detector 11, linked with the movable end of elastic member *8*. Because of dissipative losses, run up of the guided vehicle occurs up to a speed less than that of the preceding braking. At the end of run up driver *1* is engaged and this supplements the machine velocity to the nominal value.

The parameters of accumulators *6* and *7* are chosen initially from the required performance of the machine drive, the maximum velocity and moment of inertia of the moving parts of the machine.

The overall capacity of potential energy accumulators must exceed the kinetic energy of the totally loaded machine moving at its highest velocity. The capacity of accumulator *7* must considerably exceed the capacity of accumulator *6*. The parameters of the transmission mechanism *5* and accumulators *6* and *7* are chosen

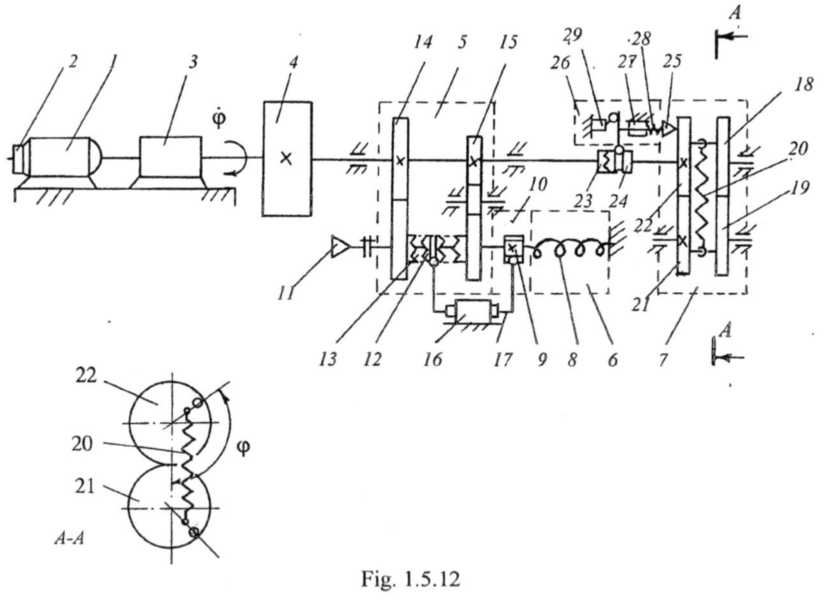

Fig. 1.5.12

so that when braking a machine moving with nominal velocity $\dot{\varphi}_n$ until it is completely at rest, accumulator *7* is fully charged to the unstable equilibrium state, and the torque from accumulator *6* applied to operating component *4* at the instant of stopping must exceed the average torque due to dry friction in the vehicle.

In the guided vehicle the energy expenditure to retain and unlock charged energy store is basically low because the stop for the second accumulator *7* is not loaded, while that of the first accumulator *6* is loaded by a torque slightly greater than that from the dry friction of the machine.

Complete relieving of the drive during the braking and speeding up phases and the reduction of dynamic impact loads in the vehicle is achieved by using potential energy accumulators, when the torque from one of these is proportional to the position of the operating component while the torque curve for the second accumulator has a distinct maximum and zero torque points.

Fig. 1.5.13 shows the cyclic advance module for a robot *(Patent USSR 1303399, IPC B25J9/00)* based on a hydraulic or pneumatic cylinder and consisting of a linear spring accumulator. Initially, arm *1* is in the end position, held by one of the latches *2* or *3* with spring *4* compressed. To start motion (for example from left to right) latch *2* is released. Bar *5* is driven rapidly forwards under the action of spring *4*, taking with it friction *6* connected to rod *7* and the manifold valve 8 which moves to the limit stop *9*. As a result, the pressurised working medium enters the left end of cylinder *10* whose right end is connected to a reservoir. The pressure in the working fluid assisting the motion of arm *1* generates work to compensate dissipative energy losses in the drive. After passing the equilibrium position arm *1* compresses spring *4* and rapidly slows down until latch *3* engages. To ensure the motion stability margin the pressure of the operating fluid is chosen so that arm *1* overruns latch *3* by a small amount. The return and seating onto latch *3* is performed by spring *4*. As a result, because of friction between *6* and bar *5*, valve *8* moves away from stop *9* closing the drainage port. This is the initial position of the drive for reverse motion of arm *1*.

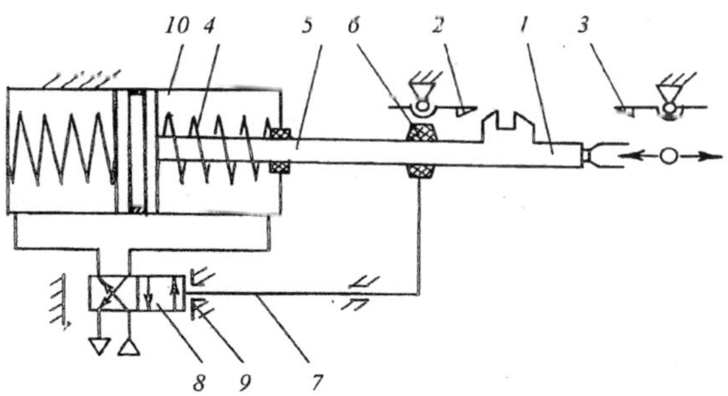

Fig. 1.5.13

1.5.6 Combining element functions in resonant robotic systems. By combining functions it is possible to reduce the number of actuators and mechanisms in drives.

Fig. 1.5.1 shows an example *(Patent USSR 1329961, IPC B25J9/00)* of the possible use of a resonant drive for the moving parts of a manufacturing machine combined with a resonant system. In the drive, the left-hand spring *5* is attached to bar *11*, constrained to move axially, and which is fitted with a roller 12 set in the curvilinear channel *13* in member *14* which moves in the transverse direction of an adjacent device. By coordinating the motion of arm *1* and member *14*, the energy in the drive of a resonant robotic system can be increased due to the displacement Δ during motion by changing the equilibrium of springs *2* and *5* in the direction of displacement of arm *1*. The value of Δ is related to the equilibrium position determined by the frictional energy losses and must be chosen to take account of the spring elasticity. The shape of curvilinear channel *13* depends on how the dissipative losses in the drive are to be compensated. A resonant drive constructed in this way eliminates the need for actuator *4*.

Superposition of the functions for the basic driver compensating dissipative losses in the drive together with the drive latch functions are provided in the manipulator module *(Patent USSR 1484685, IPC B25J9/08)* shown in Fig. 1.5.14. Initially, arm *1* is held in the extreme position by lock *2*; spring *3*, deformed by an amount δ, has one end abutting stop *4*, and the other end against plate *5* attached to the left hand end of the piston rod of cylinder *6*, which is attached to arm *1*. Plate *7*, attached to the right hand end of the piston rod, contacts the right hand stop *8*. The distance between plate *7* and lever *9* of latch *2* has the value Δ.

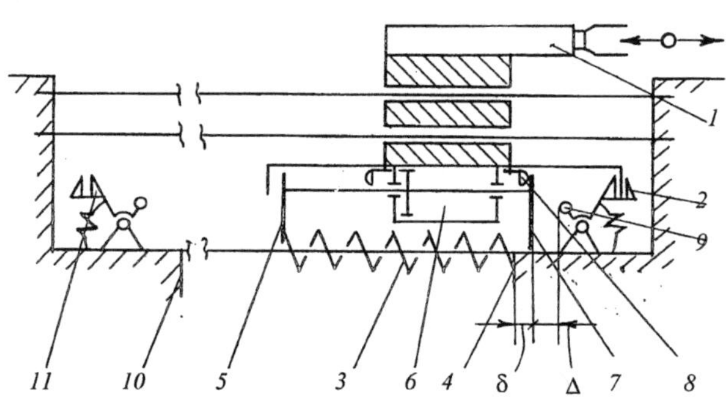

Fig. 1.5.14

Releasing and moving arm *1* occurs as follows. Operating pressure is applied to the left-hand end of cylinder *6*, which displaces the piston with its rod together with plates *5* and *7* towards the right and also deforms spring *3* by an amount Δ. Plate *7* acts on lever *9* and releases arm *1* from latch *2*. The arm begins to accelerate towards the left under the action of spring 3 whose ends act on plate 5 and stop *4*. From the instant that plate *7* passes stop *4*, spring *3* is released and arm *1* moves due to inertia. The operating pressure in cylinder *6* is cut off. Intensive braking begins at the instant when one end of spring *3* reaches the left-hand stop *10* and the other end interacts with plate *7*. When braking has finished, arm *1* is seated on the left-hand lock *11*. At this time the spring is deformed by an amount δ. The choice of stiffness for spring *3* and its working deformation δ govern the required velocity behaviour of the module. Depending on the level of energy losses in the drive the value of the additional deformation Δ of the spring is chosen taking into account the spring stiffness. In this way a single cylinder *6* can act as a means of releasing arm *1* and also to restore energy lost in friction.

The principle of superposing release functions with the basic actuator is also similarly achieved in examples with rotating robot arms *(Patent USSR 1110623, IPC B25J9/00)*.

Dynamic interrelationships of moving members are used in resonant robot arms (Fig. 1.5.15) to broaden the functional possibilities of robots *(Patent USSR 1171306, IPC B25J9/00)*. In the initial state both members *1* and *2* of a two component robot arm are at the extreme, i.e. left-hand, position. Latches *3* and *4* are locked, springs *5* and *6* contain a reserve of potential energy and actuator *7* is disengaged.

At the instant when the required motion starts, latch *3* is released and actuator *7* is engaged. Spring *5* accelerates members *1* and *2* with member *2* locked to member *1*. On a signal from a position detector, kinematically attached to member *1* latch *4* is released. As a result, member *2*, acted on by spring *6* undergoes additional acceleration. When the moving members pass the central position their kinetic energy is transformed into potential energy in the spring which causes them to brake. Near the extreme position when the members' velocity is close to zero, latches *8* and *9* are engaged, members *1* and *2* are locked and actuator 7 is

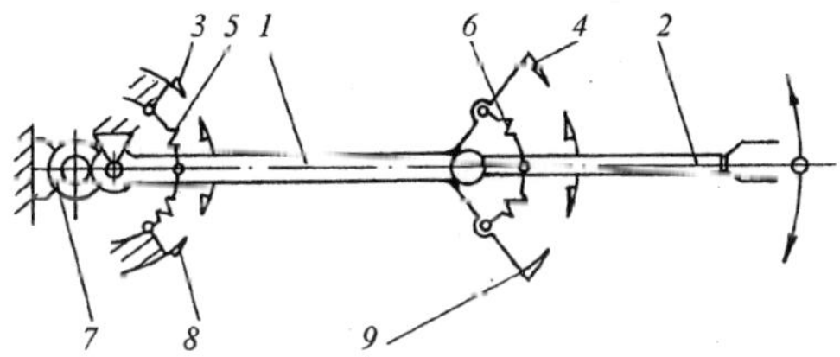

Fig. 1.5.15

de-energised. In this way the dissipation of energy due to the motion of member *2* is compensated by the dynamic interaction from member *1*. Two component construction of a robot arm can increase the number of positioning pick-up points to four.

Chapter 2. Dynamics of resonant robotic systems

§2.1 Equations of motion for resonant robots

2.1.1. In robotics, as in theoretical mechanics [78], there are two problems to be solved in robot dynamics [35, 67]. The first, which is equivalent to the inverse (first) problem of dynamics, assumes specific motions for the robot components (kinematic equations of motion); it is required to find the motive forces (either moments or direct forces), which cause the actual motion. In order to do this the dynamic forces in the drives and kinematic pairs must also be calculated. The solution of the inverse problem in dynamics makes it possible to match the choice of actuator driver and to calculate the required strength characteristics of the drive elements and robot components.

The second problem in robot dynamics, similar to the direct problem of dynamics in theoretical mechanics, consists of defining the actual motion of a manipulator (deviations from the programmed motion) due to the given parameters of the mechanical system and controls.

In the majority of cases when studying resonant robotic systems it is necessary to solve the first problem of dynamics. This arises because resonant robots were applied mainly to transport devices used in production processes to transfer products held by grippers from one position in the workspace to another, possibly including re-orientation. In this case the operating speed and positional accuracy gare of primary importance, no matter what precision in the trajectory is required. In cyclic resonant robots motion is usually provided successively by drives for each degree of freedom. As a result, the form of the gripper trajectory is completely defined by the form of the kinematic pairs (sliding or rotational), which occur in the basic kinematic scheme of the manipulator. The accuracy with which these the trajectories are followed is determined by the accuracy of manufacture of the kinematic pairs. Moreover, there is little deviation in the operating motion of resonant mechanical systems from their natural (programmed) motion if an appropriate mechanical energy store is chosen. Drive actuators play a subsidiary role in producing programmed motion, providing compensation for the energy expended on friction and to perform active work.

As a result, the problem of dynamics leads to the choice of control actions to produce the manipulator motion within defined limits or with the required behaviour. The conditions for motion can be various restraints (kinematic, geometric or energy limitations, etc.) imposed on the drive or on the manipulator elements. The provision of motion with the required behaviour can be expressed as satisfying the requirements of the motion parameters (speed, smoothness, freedom from impact, etc.) and control (energy expenditure, smoothness, heat generation in the drive, etc.).

The choice of control can be made from engineering considerations taking into account, for example, the simplicity of construction, production processes and operational advantages. In the search for motion that is stable to perturbations, the designer's choice is directed towards adaptive controls (Chapter 4). To estimate the efficiency of the chosen control it is always useful to construct optimal robot control (Chapter 3). Either the time to move the workpiece or the energy consumption in the drive can be used as a criterion of efficient functioning.

The solution of both problems in robot dynamics is based on solving the equations of motion of the robot, no matter what form they are written in. Along with the majority of methods that describe the equations of motion of a manipulator system [35, 43, 67, 91, 112, et al], we shall continue by constructing second order Lagrange's equations [3, 35, 45, 67, 78, 91].

As is well known, second order Lagrangian equations can be applied to mechanisms with a finite number of degrees of freedom and holonomous constraints. We shall assume that the mechanical model of a resonant robot represents a system of absolutely rigid bodies, connected to each other with idealised kinematic pairs. Such pairs partly limit the relative motion of adjacent components due to the imposition of ideal constraints. Constraints are called ideal if the total work of the reactions of these constraints for any virtual displacement is always zero. For a dissipative system, which a resonant robotic system represents, the second order Lagrangian equations* are written in the form

$$\frac{d}{dt}\frac{\partial T}{\partial \dot{q}_s}-\frac{\partial T}{\partial q_s}+\frac{\partial R}{\partial \dot{q}_s}=Q_s-\frac{\partial L}{\partial q_s}, \tag{2.1.1}$$

where T is the kinetic energy of the system, S is the number of generalised coordinates q, which is equal to the number of degrees of freedom, $\dot{q}_s$ are generalised velocities, R is Rayleigh's dissipative function, Q_s are generalised forces, L is the potential energy of the system. The kinetic energy of the system T consists of the kinetic energy of the components and elements of the drives that move within the manipulator.

For the analysis of the dynamics of resonant systems it is necessary to take into account the dissipative forces. The application of Lagrange's equations assumes that friction forces are external and do not depend on the reactions in the kinematic pairs. As shown in §1.3, the dissipative forces in the majority of cases are proportional to the velocities $\dot{q}_s$. Rayleigh's dissipative function is given by the formula

$$R=\frac{1}{2}\sum_{i,k=1}^{s} b_{ik}\dot{q}_i\dot{q}_k, \tag{2.1.2}$$

where $b_{i,k}$ are constant coefficients, and in resonant systems appear as positive definite square form of the generalised velocities. Generalised dissipative forces are described by the formula

* Below, we shall refer to these equations simply as "Lagrange's equations," since in this book, Lagrange's equations of the first order are not considered.

$$\Phi_s = -\partial R / \partial \dot{q}_s \,. \tag{2.1.3}$$

The generalised forces Q_s are found from the conditions that the elementary work of these forces on the allowable variations δq_s in generalised coordinates, and the work δA of the active forces F_i applied to the components of the mechanism with virtual displacements δr_i must be equal. Remembering from [78] that the virtual displacement represents the displacement of points in the system from one possible position at one instant, to another extremely close possible position at the same instant of time t. In practice, to find Q_s, the elementary work corresponding to the change in one generalised coordinate q_s of the active force for a virtual displacement is calculated, assuming that the remaining generalised coordinates are fixed. As a result the coefficient with q_s in the work equation is equal to Q_s.

$$\delta A = \sum_{i=1}^{N} F_i \delta r_i = \sum_{s=1}^{n} Q_s \delta q_s \,. \tag{2.1.4}$$

For a generalised linear coordinate the generalised force has the dimension of force, but for a generalised angular coordinate the generalised force has the dimension of a moment. We shall consider the control forces created by the actuator driver as the active forces F_i in a resonant robot, together with gravity forces of the manipulator elements and the forces resisting motion of the gripper. The friction forces in the robot will also be considered as active but they are taken into account in the Lagrangian equations by the Rayleigh dissipative function of equation (2.1.2).

In resonant manipulator systems, the forces that depend only on the generalised coordinate are conveniently taken into account through changes in the potential energy L of the system. In the given case, the potential forces are the elastic forces in springs, which have been studied in detail in §1.4.

In order to write down Lagrange's equations, we shall make use of the so-called *Lagrangian formalism* that describes consecutive actions, and makes it possible to write down the equations of motion for any system of coordinates as a standard example [3].

1. Choose the independent (generalised) coordinates $q_1, \ldots, q_n$, which will be used to write the equations.
2. Find the generalised forces Q_s. If the active forces F_i are functions of the coordinate points of the system or their velocities, then when calculating the generalised forces, the generalised coordinates q_s and their derivatives $\dot{q}_s$ must be used to express the Cartesian coordinates of the points and their derivatives.
3. Calculate the kinetic energy of the system as a function of the coordinates q_s, their derivatives $\dot{q}_s$, and perhaps, time t.
4. Calculate the potential energy L of the system as a function of the generalised coordinates q_s, and Rayleigh's dissipative function as a

function of the generalised velocities $\dot{q}_s$.

5. Generate the full and partial derivatives shown in equation (2.1.2), i.e. substitute the values found above for forces and energy into the Lagrangian equations.

2.1.2. As an example we shall write the equations of motion in Lagrangian form for the *MARS resonant robot* [4], whose kinematic scheme is shown in Fig.2.1.1.

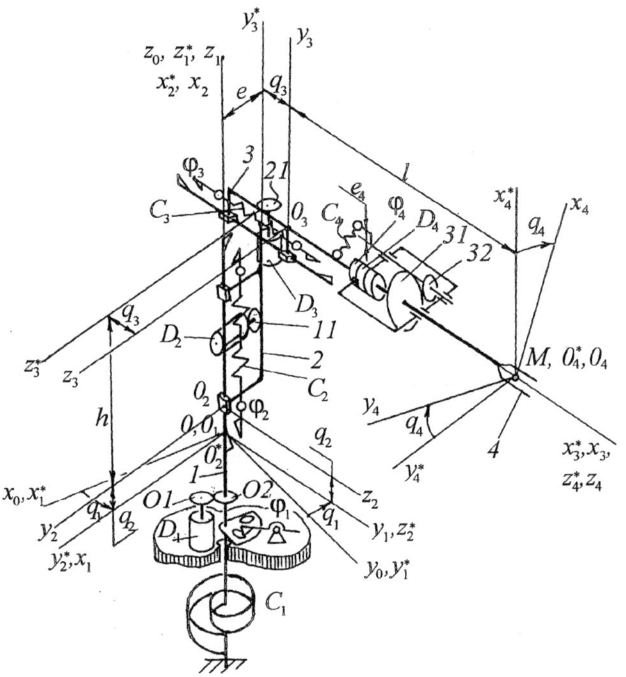

Fig. 2.1.1

The MARS resonant robot is intended for automatically loading and unloading small components in high-speed manufacturing environments, for example, as a production accessory in welding and in presses for small components. The robot operates in a cylindrical coordinate system and has four degrees of freedom. The basic technical data for the robot are shown in the table of §1.2. By using modular construction (three modules: rotation, elevation and telescopic operation of the robot arm) and operating in a cyclic manner with locking at the extreme positions, high positioning accuracy of the robot gripper is possible - 0.05mm. The resonant arrangement of the electro-mechanical drive and the use of adaptive control ensure reliability and high-speed operation of the robot.

MARS is arranged as follows:

The column *1* rotates about the vertical axis under the action of spiral spring with stiffness C_1 and driver D_1 attached to the column through a single-stage reduction gear with gears O_1 and O_2. The controlling latch Φ_1 holds the column at the positioning point. Lifting carriage *2* is attached to column *1*. The lift drive consists of driver D_2 with driving gear *11*, a linear spring with stiffness C_2 and the control latch Φ_2.

The module to produce horizontal motion of arm *3* is attached to the driven component of carriage *2*. Its drive consists of driver D_3 with gear *21*, linear spring with stiffness C_3 and latch Φ_3.

The module to rotate gripper *4* around the axis of arm *3* of the robot has a drive consisting of a motor and speed-reducer D_4 with built-in brake Φ_4. The body of the motor and speed-reducer is rigidly attached to arm *3* and the output shaft is connected to gripper *4* through gear *31*. In addition, gear *31* meshes with gear *32* attached to crank *33*, the end of which is attached by a spring with stiffness C_4 to arm *3*. Spring C_4 operates in a manner similar to that shown in Fig. 1.4.4, *d*.

In the discussion below we shall use conventional meanings [67].

The direction cosine matrix

$$A_{s-1,s} = \begin{bmatrix} \cos(x_{s-1}, x_s) & \cos(x_{s-1}, y_s) & \cos(x_{s-1}, z_s) \\ \cos(y_{s-1}, x_s) & \cos(y_{s-1}, y_s) & \cos(y_{s-1}, z_s) \\ \cos(z_{s-1}, x_s) & \cos(z_{s-1}, y_s) & \cos(z_{s-1}, z_s) \end{bmatrix}$$

together with the radius-vector $r_{0_s}^{s-1} = \left(x_{0_s}^{s-1},\ y_{0_s}^{s-1},\ z_{0_s}^{s-1}\right)^T$ from the origin 0_s projected onto the axes of the $(s-1)th$ coordinate system, give the unique position of the $0_s x_s y_s z_s$ system of coordinates relative to the $0_{s-1} x_{s-1} y_{s-1} z_{s-1}$ (Fig. 2.1.2). As a result the following type of relationships occur

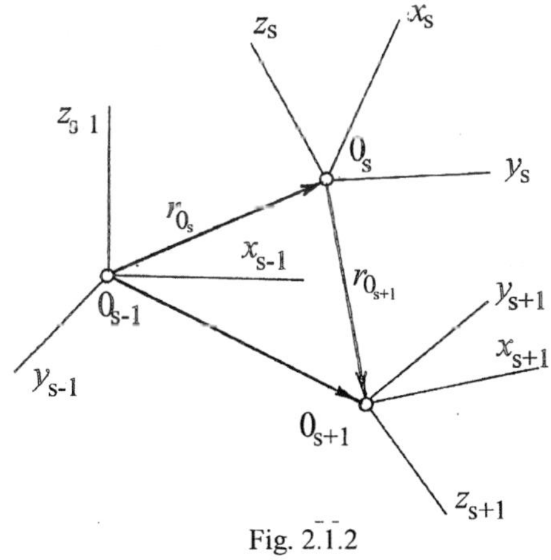

Fig. 2.1.2

$$A_{s-1,s+1} = A_{s-1,s} A_{s,s+1} \quad ,$$
$$r^{s-1}_{0_{s+1}} = r^{s-1}_{0_s} + A_{s-1,s} r^{s}_{0_{s+1}} \quad , \tag{2.1.5}$$

where the upper index shows the system of coordinates onto which the vector is projected.

In the description of the kinematics of manipulators it is convenient to use the four-dimensional column vector $R^{s-1}_{0_s} = \left(r^{s-1}_{0_s},\ 1\right)^T$ and the matrix $H_{s-1,s}$ for transformation from the $s-th$ coordinate system to the $(s-1)th$ [67, 91]

$$H_{s-1,s} = \begin{bmatrix} A_{s-1,s} & r^{s-1}_{0_s} \\ 0 & 1 \end{bmatrix}.$$

For this to be valid, the expressions must be of the form

$$R^{s-1}_{0_s} = H_{s-1,s} R^{s}_{0_s},$$
$$H_{s-1,s+1} = H_{s-1,s} H_{s,s+1}. \tag{2.1.4}$$

Continuing the Lagrangian formulation we now choose the following independent (generalised) coordinates:

q_1 - the angle of rotation about the vertical axis;

q_2 - the vertical displacement of the carriage relative to the column;

q_3 - the advance of the arm relative to the carriage;

q_4 - the angle of rotation of the gripper about the axis of the robot arm.

We shall find the position and velocity functions for the robot elements. To do this the following coordinates system are chosen.

The $0_0 x_0 y_0 z_0$ system is that of the body. The z_0 – axis is the rotational axis of the column, and the x_0 – and y_0 – axes are perpendicular to the z_0 – axis.

The $0_1 x_1 y_1 z_1$ system is that of the column. The z_1 – axis is directed along the rotation axis to coincide with the z_0 – axis; the x_1 – and y_1 – axes are perpendicular to the z_1 – axis. The initial position of the first system which has coordinates $0^*_1 x^*_1 y^*_1 z^*_1$ (the position for which $q_1 = 0$) is chosen to coincide with the $0_0 x_0 y_0 z_0$ system when the direction cosine matrix connecting these systems is the unit matrix $A_{0,1*} = \begin{bmatrix} 1 & 0 & 0 \\ 0 & 1 & 0 \\ 0 & 0 & 1 \end{bmatrix}$, the radius vector $r^0_{0_1} = 0$ and the direction cosine matrix connecting the $0_0 x_0 y_0 z_0$ system with the $0_1 x_1 y_1 z_1$ system has the form

$$A_{0,1} = A_{0,1*} A_{1*,1} = \begin{bmatrix} \cos q_1 & -\sin q_1 & 0 \\ \sin q_1 & \cos q_1 & 0 \\ 0 & 0 & 1 \end{bmatrix}. \tag{2.1.7}$$

Eventually, we obtain the transformation matrix from the $0_1 x_1 y_1 z_1$ system to the $0_0 x_0 y_0 z_0$ system in the form

$$H_{0,1} = \begin{bmatrix} A_{0,1} & r_{0_1}^0 \\ 0 & 1 \end{bmatrix} = \begin{bmatrix} \cos q_1 & -\sin q_1 & 0 & 0 \\ \sin q_1 & \cos q_1 & 0 & 0 \\ 0 & 0 & 1 & 0 \\ 0 & 0 & 0 & 1 \end{bmatrix}. \qquad (2.1.8)$$

The system of coordinates connected with the carriage is arranged so that the x_2 – axis is directed along the z_1 – axis of column *1*. The initial position of the $0_2^* x_2^* y_2^* z_2^*$ system is such that 0_2^* coincides with 0 and 0_1, but the y_2^* – and z_2^* – axes are correspondingly arranged along the x_1 – and y_1 – axes. Taking into account that $r_{0_2^*}^0 = 0$, we obtain

$$A_{1,2^*} = A_{1,2} = \begin{bmatrix} 0 & 1 & 0 \\ 0 & 0 & 1 \\ 1 & 0 & 0 \end{bmatrix},$$

$$H_{1,2} = \begin{bmatrix} 0 & 1 & 0 & 0 \\ 0 & 0 & 1 & 0 \\ 1 & 0 & 0 & q_2 \\ 0 & 0 & 0 & 1 \end{bmatrix}. \qquad (2.1.9)$$

The coordinate system for the robot arm is such that the x_3 – axis coincides with the axis of the arm and is simultaneously parallel with the y_1 and z_2 axes. The $0_3^* x_3^* y_3^* z_3^*$ coordinate system is chosen so that $r_{0_3^*}^2 = (h, \ -e, \ 0)^T$ and the x_3–, y_3 – and z_3 – axes are respectively parallel to the z_2^*–, x_2^* – and y_2^* – axes. Such choices stipulate the following equations

$$A_{2,3^*} = A_{2,3} = \begin{bmatrix} 0 & 1 & 0 \\ 0 & 0 & 1 \\ 1 & 0 & 0 \end{bmatrix},$$

$$r_{0_3}^2 = \begin{bmatrix} h \\ -e \\ q_3 \end{bmatrix}, \qquad (2.1.10)$$

$$H_{2,3} = \begin{bmatrix} A_{2,3} & r_{0_3}^0 \\ 0 & 1 \end{bmatrix}.$$

The origin of the $0_4 x_4 y_4 z_4$ coordinate system for the robot gripper is placed at the position M of the gripper with coordinate $r_{0_4^*}^3 = (l, \ 0, \ 0)^T$. The x_4^*–, y_4^* – and z_4^* – axes are arranged respectively parallel to the y_3–, z_3 – and x_3 – axes. As a result, the transformation function takes the form

$$H_{3,4} = \begin{bmatrix} 0 & 0 & 1 & l \\ \cos q_4 & \sin q_4 & 0 & 0 \\ -\sin q_4 & \cos q_4 & 0 & 0 \\ 0 & 0 & 0 & 1 \end{bmatrix}. \tag{2.1.11}$$

We shall now find the position functions for the gripper.

The expansion of the radius vector of position M of the gripper is found from the equation

$$R_M^0 = H_{0,4} R_M^4 , \tag{2.1.12}$$

where $R_M^4 = (0, \quad 0, \quad 0, \quad 1)^T$, and the transformation function is found from the expression

$$H_{0,4} = H_{0,1} H_{1,2} H_{2,3} H_{3,4} . \tag{2.1.13}$$

After substituting (2.1.8), (2.1.9), (2.1.10) and (2.1.11) into (2.1.12) and (2.1.13) we obtain

$$R_M^0 = \begin{bmatrix} -(l+q_3)\sin q_1 - e\cos q_1 \\ (l+q_3)\cos q_1 - e\sin q_1 \\ h+q_2 \\ 1 \end{bmatrix} \tag{2.1.14}$$

$$H_{0,4} = \begin{bmatrix} A_{0,4} & r_M^0 \\ 0 & 1 \end{bmatrix} =$$

$$= \begin{bmatrix} -\cos q_1 \sin q_4 & \cos q_1 \cos q_4 & -\sin q_1 & -(l+q_3)\sin q_1 - e\cos q_1 \\ -\sin q_1 \sin q_4 & \sin q_1 \cos q_4 & \cos q_1 & (l+q_3)\cos q_1 - e\sin q_1 \\ \cos q_4 & \sin q_4 & 0 & h+q_2 \\ 0 & 0 & 0 & 1 \end{bmatrix}, \tag{2.1.15}$$

To project the velocity of position M of the gripper onto the $0_0 x_0 y_0 z_0$ axes we find the derivative of the column vector r_M^0 (2.1.15) with respect to time

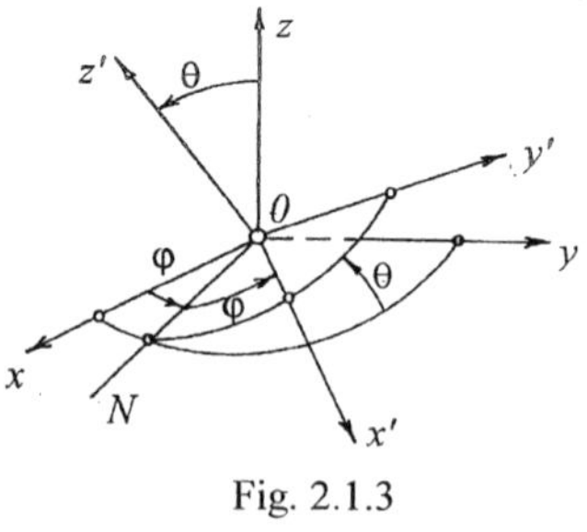

Fig. 2.1.3

$$V^0_{Mx} = \dot{q}_1[e\sin q_1 - (l+q_3)\cos q_1] - \dot{q}_3 \sin q_1,$$
$$V^0_{My} = -\dot{q}_1[e\cos q_1 - (l+q_3)\sin q_1] - \dot{q}_3 \cos q_1, \qquad (2.1.16)$$
$$V^0_{Mz} = \dot{q}_2.$$

The position functions of the gripper with position M can be written as the three linear coordinates $x_M = x_M(q_n)$, $y_M = y_M(q_n)$, $z_M = z_M(q_n)$ obtained in (2.1.14) and the three *Euler angles* $\psi = \psi(q_n)$, $\theta = \theta(q_n)$, $\varphi = \varphi(q_n)$, where $n = 1,\ldots,4$.

The Euler angles [78] are chosen as shown in Fig. 2.1.3, where the coordinate system $0x'y'z'$ is related to the motion of the moving body relative to the stationary $0xyz$ system. The nodal line 0N is the intersection of the $x0y$ and $x'0y'$ planes. The *precession* angle, ψ, is shown on plane $x0y$ as the nodal line relative to the $0x$ – axis. The positive direction of angle ψ is measured from the $0x$ – axis to the 0N – axis in the clockwise direction when viewed from the $0z$ – axis. The *nutation* angle θ is in the $z0z'$ – plane and is measured from the $0z$ – axis to the $0z'$ – axis in the anti-clockwise direction when viewed from the 0N – axis. The *simple rotation* angle φ is in the $x'0y'$ – plane and is measured from the nodal line 0N to the $0x'$ – axis in the positive direction.

We shall use the well-known relationships [78]

$$\cos\theta = \cos(z_0\hat{\ }z_4), \qquad \sin\psi = \cos(x_0\hat{\ }z_4)/\sin\theta,$$
$$\cos\psi = -\cos(y_0\hat{\ }z_4)/\sin\theta, \qquad \cos\varphi = \cos(z_0\hat{\ }y_4)/\sin\theta, \qquad (2.1.17)$$
$$\sin\theta = (1-\cos^2\theta)^{1/2}, \qquad \sin\varphi = \cos(z_0\hat{\ }x_4)/\sin\theta,$$

and expressing the Euler angles through the elements of matrix $A_{0,4}$ from (2.1.15)

$$\cos\theta = 0,\ \sin\theta = 1,\ \cos\varphi = \sin q_4,\ \sin\varphi = \cos q_4,\ \sin\psi = -\sin q_1,\ \cos\psi = -\cos q_1,$$

from which it follows that

$$\theta = \pi/2, \quad \varphi = \pi/2 - q_4, \quad \psi = \pi + q_1. \qquad (2.1.18)$$

2.1.3. We shall now find the velocity of points and components of the robot. Absolute angular velocities will be defined by the recurrence formula [67]

$$\bar{\Omega}^{i+1}_{i+1} = A^T_{i,i+1}\bar{\Omega}^i_i + \bar{\omega}^{i+1}_{i+1}, \qquad (2.1.19)$$

where $\bar{\omega}^{i+1}_{i+1}$ is the vector of the angular velocity of the $(i+1)$ th component relative to the i th in the $(i+1)$th – coordinate system, $\bar{\Omega}^{i+1}_{i+1}$ and $\bar{\Omega}^i_i$ are the vectors of the absolute angular velocities of the $(i+1)$ th and i th components in the $(i+1)$ th and i th coordinate systems respectively, $A^T_{i,i+1}$ is the transpose of the direction cosine matrix of the $(i+1)$ th coordinate system relative to the i th.

For the column *1*, the carriage *2*, arm *3* and gripper *4*, taking into account that $\bar{\Omega}^0_0 = \bar{\omega}^2_2 = \bar{\omega}^3_3 = 0$, $\bar{\omega}^1_1 = (0,\ 0,\ \dot{q}_1)^T$, $\bar{\omega}^4_4 = (0,\ 0,\ \dot{q}_4)^T$, we can write

$$\bar{\Omega}^1_1 = A^T_{0,1}\bar{\Omega}^0_0 + \bar{\omega}^1_1 = (0,\ 0,\ \dot{q}_1)^T, \qquad (2.1.20)$$

$$\bar{\Omega}^2_2 = A^T_{1,2}\bar{\Omega}^1_1 = (\dot{q}_1,\ 0,\ 0)^T, \qquad (2.1.21)$$

$$\bar{\Omega}_3^3 = A_{2,3}^T \bar{\Omega}_2^2 = (0, \quad \dot{q}_1, \quad 0)^T, \tag{2.1.22}$$

$$\bar{\Omega}_4^4 = A_{3,4}^T \bar{\Omega}_3^3 + \bar{\omega}_4^4 = \begin{bmatrix} \dot{q}_1 \cos q_4 \\ \dot{q}_1 \sin q_4 \\ \dot{q}_4 \end{bmatrix}. \tag{2.1.23}$$

The calculation of linear velocities of the coordinate systems associated with the moving components will be derived from the recurrence formula [76]

$$V_{0_{s+1}}^s = V_{0_s}^s + \bar{\Omega}_s^s \times r_{0_{s+1}}^s + \dot{r}_{0_{s+1}}^s, \quad V_{0_{s+1}}^{s+1} = A_{s,s+1}^T V_{0_{s+1}}^s, \tag{2.1.24}$$

which makes it possible to calculate the velocity of an arbitrary point 0_{s+1} of the

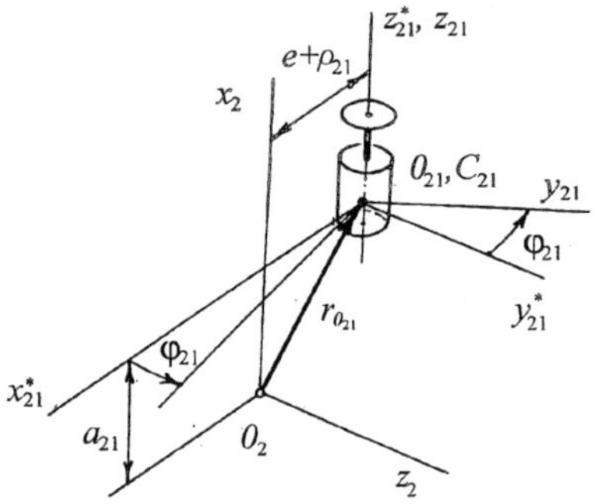

Fig. 2.1.4

moving coordinate system projected onto the axes of the sth and $(s+1)$th coordinate systems.

The velocities of points $0_1, \ldots, 0_4$ are found from the expressions

$$V_{0_1}^1 = V_{0_1}^0 = 0, \quad V_{0_2}^2 = (\dot{q}_2, \quad 0, \quad 0)^T, \quad V_{0_3}^3 = \begin{bmatrix} -e\dot{q}_1 + \dot{q}_3 \\ \dot{q}_2 \\ -\dot{q}_3 \dot{q}_4 \end{bmatrix},$$

$$V_{0_4}^4 = \begin{bmatrix} \dot{q}_2 \cos q_4 + (l + q_3)\dot{q}_1 \sin q_4 \\ \dot{q}_2 \sin q_4 - (l + q_3)\dot{q}_1 \cos q_4 \\ -e\dot{q}_1 + \dot{q}_3 \end{bmatrix} \tag{2.1.25}$$

The analysis of the velocities of the moving elements of the robot drive is made using formulas (2.1.24) and (2.1.19). Thus, superposing the local coordinate system $0_{21}x_{21}y_{21}z_{21}$ associated with both the gearwheel *21* and the rotor of the driver D_3 on their centre of mass at C_{21} (Fig. 2.1.4), the point 0_{21} will have coordinates $r_{0_{21}}^2 = r_{C_{21}}^2 = (a_{21}, \quad -e - \rho_{21}, \quad 0)^T$, where ρ_{21} is the radius of gear *21*. Then the equations for finding the velocities of the component *21* take the form

$$V_{0_{21}}^{2} = V_{0_2}^{2} + \bar{\Omega}_2^2 \times r_{0_{21}}^{2} + \dot{r}_{0_{21}}^{2} = \begin{bmatrix} \dot{q}_2 \\ 0 \\ -(e+\rho_{21})\dot{q}_1 \end{bmatrix},$$

$$V_{0_{21}}^{21} = V_{C_{21}}^{21} = A_{2,21}^{T} V_{0_{21}}^{2} = \begin{bmatrix} -(e+\rho_{21})\dot{q}_1 \cos q_3 / \rho_{21} \\ -(e+\rho_{21})\dot{q}_1 \sin q_3 / \rho_{21} \\ \dot{q}_2 \end{bmatrix}, \tag{2.1.26}$$

$$\bar{\Omega}_{21}^{21} = A_{2,21}^{T}\bar{\Omega}_2^2 + \bar{\omega}_{21}^{21} = \left[\dot{q}_1,\ 0,\ \dot{q}_3/\rho_{21}\right]^T,$$

where $V_0^2 = (\dot{q}_1,\ 0,\ 0)^T$, $\bar{\Omega}_2^2 = (\dot{q}_1,\ 0,\ 0)^T$, $\dot{r}_{0_{21}}^2 = 0$,

$$\bar{\omega}_{21}^{21} = (0,\ 0,\ \dot{q}_3/\rho_{21})^T,\quad A_{2,21} = \begin{bmatrix} 0 & 0 & 1 \\ \cos q_3/\rho_{21} & -\sin q_3/\rho_{21} & 0 \\ \sin q_3/\rho_{21} & \cos q_3/\rho_{21} & 0 \end{bmatrix}.$$

Similarly, we find the velocities of components *11*, *31* and *32*

$$V_{0_{11}}^{11} = \dot{q}_1 \begin{bmatrix} a_{11}\sin\varphi_{11} \\ -a_{11}\cos\varphi_{11} \\ -b_{11} \end{bmatrix},\quad \bar{\Omega}_{11}^{11} = \begin{bmatrix} \dot{q}_1\cos\varphi_{11} \\ \dot{q}_1\sin\varphi_{11} \\ \dot{q}_2/\rho_{11} \end{bmatrix},$$

$$V_{0_{31}}^{31} = \begin{bmatrix} \dot{q}_2\cos q_4 + (a_{31}+q_3)\dot{q}_1\sin q_4 \\ \dot{q}_2\sin q_4 - (a_{31}+q_3)\dot{q}_1\cos q_4 \\ -e\dot{q}_1 + \dot{q}_3 \end{bmatrix},\quad \bar{\Omega}_{31}^{31} = \begin{bmatrix} \dot{q}_1\cos q_4 \\ \dot{q}_1\sin q_4 \\ \dot{q}_4 \end{bmatrix}, \tag{2.1.27}$$

$$V_{0_{32}}^{32} = \begin{bmatrix} \dot{q}_2\cos\varphi_{32} - (a_{32}+q_3)\dot{q}_1\sin\varphi_{32} \\ -\dot{q}_2\sin\varphi_{32} - (a_{32}+q_3)\dot{q}_1\cos\varphi_{32} \\ -(e+d_{32})\dot{q}_1 + \dot{q}_3 \end{bmatrix},\quad \bar{\Omega}_{32}^{32} = \begin{bmatrix} \dot{q}_1\cos\varphi_{32} \\ -\dot{q}_1\sin\varphi_{32} \\ \dot{q}_4\rho_{31}/\rho_{32} \end{bmatrix},$$

where $r_{0_{11}}^{1} = (-a_{11},\ b_{11},\ d_{11})^T$, $\varphi_{11} = q_2/\rho_{11}$, $\bar{\omega}_{11}^{11} = (0,\ 0,\ \dot{q}_2/\rho_{11})^T$, $\bar{\omega}_{31}^{31} = (0,\ 0,\ \dot{q}_4)^T$, $r_{0_{31}}^{31} = (a_{31},\ 0,\ 0)^T$, $\bar{\omega}_{32}^{32} = (0,\ 0,\ \dot{q}_4\rho_{31}/\rho_{32})^T$, $r_{0_{32}}^{31} = (a_{32},\ 0,\ -d_{32})^T$, $\varphi_{32} = q_4\rho_{31}/\rho_{32}$, ρ_{31} and ρ_{32} are the radii of gears *31* and *32* respectively.

Note that the origin 0_i of the coordinate system for components *11*, *21*, *31* and *32* is placed at the centre of mass C_i of these components.

2.1.4. The generalised forces Q_s will be defined according to (2.1.4) by the formula [67]

$$Q_s = \frac{\delta A_s}{\delta q_s} = \sum_{k=s}^{n} G_k \frac{\partial r_{C_k}^s}{\partial q_s} + \sum_{k=s}^{n}\sum_{l=1}^{m_k} G_{kl}\frac{\partial r_{C_{kl}}^s}{\partial q_s} + D_s \frac{\partial \varphi_{D_s}}{\partial q_s}, \tag{2.1.28}$$

where G_k and G_{kl} are the gravity forces due to the operating components of the manipulator and drive elements, $r_{C_k}^s$ and $r_{C_{kl}}^s$ are the radius vectors of the centres of mass of the manipulator components and drive elements projected onto the axes of the $s-$ th coordinate system which change as the generalised coordinate q_s varies, and D_s and φ_{D_s} are the torque and angle of rotation of the driver which causes motion through coordinate q_s.

Then, for the four generalised coordinates we can write four generalised forces

$$\begin{aligned} &Q_1 = D_1\rho_{02}/\rho_{01}, \\ &Q_2 = D_2/\rho_{11} - (m_2 + m_3 + m_4 + m_{21} + m_{31} + m_{32})g, \\ &Q_3 = D_3/\rho_{21}, \\ &Q_4 = D_4. \end{aligned} \tag{2.1.29}$$

where ρ_{01}, ρ_{02}, ρ_{11} and ρ_{21} are the radii of the respective gears.

2.1.5. Using the previously calculated values for the velocities of the manipulator components and the drive elements, we can write the kinetic energy of the robot as the sum of the kinetic energies of all the components in terms of the generalised coordinates $q_1, \ldots, q_4$ and velocities $\dot{q}_1, \ldots, \dot{q}_4$.

The kinetic energy of the $i-$ th component undergoing spatial motion is equal to the sum of the kinetic energy in translation along the trajectory of the mass centre with velocity V_{C_i} and the kinetic energy due to spherical motion around the mass centre C_i, and is given by

$$\begin{aligned} T_i = (m_i V_{C_i}^2 + J_{x_i}\Omega_{x_i}^2 + J_{y_i}\Omega_{y_i}^2 + J_{z_i}\Omega_{z_i}^2)/2 - \\ - J_{x_i y_i}\Omega_{x_i}\Omega_{y_i} - J_{y_i z_i}\Omega_{y_i}\Omega_{z_i} - J_{z_i x_i}\Omega_{z_i}\Omega_{x_i}. \end{aligned} \tag{2.1.30}$$

Here $x_i y_i z_i$ is the coordinate system for the component with centre of mass C_i; J_{x_i}, J_{y_i}, J_{z_i} are the mass moments of inertia of the component about its own axes; $J_{x_i y_i}$, $J_{y_i z_i}$, $J_{z_i x_i}$ are the mass products of inertia; Ω_{x_i}, Ω_{y_i}, Ω_{z_i} are the projections of the instantaneous angular velocities on the x_i, y_i, z_i axes. If the coordinate axes are chosen to be the principal axes of inertia, then the mass products of inertia are zero, i.e. $J_{x_i y_i} = J_{y_i z_i} = J_{z_i x_i} = 0$.

Using the values for velocities found earlier in (2.1.20) to (2.1.27), we can write the equations for kinetic energy of the robot components as

$$\begin{aligned} &T_1 = J_{z1}\dot{q}_1^2/2, \qquad T_2 = (m_2\dot{q}_2^2 + J_{x2}\dot{q}_1^2)/2, \\ &T_3 = m_3[(e^2 + q_3^2)\dot{q}_1^2 + \dot{q}_2^2 + \dot{q}_3^2 - 2e\dot{q}_1\dot{q}_3]/2 + J_{y3}\dot{q}_1^2/2, \\ &T_4 = [(J_{x4}\cos^2 q_4 + J_{y4}\sin^2 q_4)\dot{q}_1^2 + J_{z4}\dot{q}_4^2]/2 + \\ &\qquad + m_4\{\dot{q}_1^2[e^2 + (l + q_3)^2] + \dot{q}_2^2 + \dot{q}_3^2 - 2e\dot{q}_1\dot{q}_3\}/2, \\ &T_{11} = m_{11}(a_{11}^2 + b_{11}^2)\dot{q}_1^2/2 + (J_{x11}\cos^2 q_2/\rho_{11} + J_{y11}\sin^2 q_2/\rho_{11})\dot{q}_1^2/2 + \\ &\qquad + J_{z11}\dot{q}_2^2/2\rho_{11}^2, \\ &T_{21} = m_{21}[(e + \rho_{21})^2\dot{q}_1^2 + \dot{q}_2^2]/2 + (J_{x21}\dot{q}_1^2 + J_{z21}\dot{q}_3^2/\rho_{21}^2)/2, \end{aligned} \tag{2.1.31, a}$$

$$T_{31} = m_{31}\{\dot q_2^2 + [(a_{31}+q_3)^2 + e^2]\dot q_1^2 + \dot q_3^2 - 2e\dot q_1 \dot q_3\}/2 +$$
$$+ (J_{x31}\cos^2 q_4 + J_{y31}\sin^2 q_4)\dot q_1^2/2 + J_{z31}\dot q_4^2/2,$$
$$T_{32} = m_{32}\{\dot q_2^2 + [(a_{32}+q_3)^2 + (e^2+d_{32})^2]\dot q_1^2 + \dot q_3^2 - 2(e+d_{32})\dot q_1 \dot q_3\}/2 + \quad (2.1.31,\text{b})$$
$$+ (J_{x32}\cos^2 q_4 \rho_{31}/\rho_{32} + J_{y32}\sin^2 q_4 \rho_{31}/\rho_{32})\dot q_1^2/2 +$$
$$+ J_{z32}(\dot q_4^2 \rho_{31}/\rho_{32})^2/2.$$

Assuming that the total kinetic energy of the robot is of quadratic form in $\dot q_s$ with coefficients a_{ik} which depend on q_s and which are the elements of a symmetric matrix $A = (a_{ik})_1^4$, we can write the following equation:

$$T = \tfrac{1}{2}\sum_{i,k=1}^{4} a_{ik}(q_1, q_2, q_3, q_4)\dot q_i \dot q_k = \tag{2.1.32}$$
$$= \{[a_{11}^0 + a_{11}(q_2, q_3, q_4)]\dot q_1^2 + a_{22}^0\dot q_2^2 + a_{33}^0\dot q_3^2 + a_{44}^0\dot q_4^2 + 2a_{13}^0\dot q_1\dot q_3\}/2,$$

where the terms a_{ik}^0 which are constants, and the terms a_{ik} are variables given by

$$a_{11}^0 = J_{z1} + J_{x2} + J_{y3} + J_{x21} + (m_3 + m_4 + m_{31})e^2 + m_{11}(a_{11}^2 + b_{11}^2) +$$
$$+ m_{21}(e+\rho_{21})^2 + m_4 l^2 + m_{31}a_{31}^2 + m_{32}[a_{32}^2 + (e+d_{32})^2],$$
$$a_{11}(q_2, q_3, q_4) = (m_3 + m_4 + m_{31} + m_{32})q_3^2 + 2(lm_4 + a_{31}m_{31} + a_{32}m_{32})q_3 +$$
$$+ (J_{x4} + J_{x31})\cos^2 q_4 + (J_{y4} + J_{y31})\sin^2 q_4 +$$
$$+ J_{x11}\cos^2 q_2/\rho_{11} + J_{y11}\sin^2 q_2/\rho_{11} + J_{x32}\cos^2 q_4\rho_{31}/\rho_{32} + \tag{2.1.33}$$
$$+ J_{y32}\sin^2 q_4 \rho_{31}/\rho_{32},$$
$$a_{22}^0 = m_2 + m_3 + m_4 + m_{21} + J_{z21}/\rho_{21}^2,$$
$$a_{33}^0 = m_3 + m_4 + m_{31} + m_{32} + J_{z21}/\rho_{21}^2,$$
$$a_{44}^0 = J_{z4} + J_{z31} + J_{z32}\rho_{31}^3/\rho_{32}^2,$$
$$a_{13}^0 = -(m_3 + m_4 + m_{31} + m_{32}) - d_{32}m_{32}.$$

2.1.6. According to (2.1.2), and assuming that the matrix $B = diag\{b_1, \ldots, b_4\}$, the Rayleigh dissipative function can be written as

$$R = \tfrac{1}{2}\sum_{i=1}^{4} b_i \dot q_i^2 = (b_1\dot q_1^2 + b_2\dot q_2^2 + b_3\dot q_3^2 + b_4\dot q_4^2)/2\,. \tag{2.1.34}$$

The potential energy of the robot due to the deformation energy of the four springs takes the form

$$L = \sum_{i=1}^{4} L_i = \tfrac{1}{2}\sum_{i=1}^{4} C_i q_i^2 =$$
$$= C_1 q_1^2/2 + C_2(q_2 - A_2)^2/2 + C_3 q_3^2/2 + \tag{2.1.35}$$
$$+ C_4\{[e_4^2 + (\rho_{31}+\rho_{32})^2 + 2e_4(\rho_{31}+\rho_{32})\cos q_4 \rho_{31}/\rho_{32}]^{1/2} - l_4\}^2/2,$$

where l_1 and l_4 are the initial lengths of the springs, C_i is the spring stiffness constant, A_2 is the half travel of the carriage *2* (see Fig. 2.1.1), e_4 is the eccentricity of the crank in the rotational drive for the robot gripper.

2.1.7. We can now write Lagrange's equation (2.1.1), taking into account (2.1.2), (2.1.28), (2.1.32), (2.1.34) and (2.1.35), in the following form

$$\sum_{k+1}^{4} a_{sk}(q_1,q_2,q_3,q_4)\ddot{q}_k + \sum_{i,k=1}^{4} \Gamma_{ik;s}\dot{q}_i\dot{q}_k + \sum_{i=1}^{4} b_i\dot{q}_i = = Q_i - \sum_{i=1}^{4} C_i q_i, \tag{2.1.36}$$

where $\Gamma_{ik;s} = [i \quad k; \quad s] = \left(\frac{\partial a_{is}}{\partial q_k} + \frac{\partial a_{ks}}{\partial q_i} - \frac{\partial a_{ik}}{\partial q_s}\right)/2$ is the first order Christoffel symbol [68]

From (2.1.33), we obtain

$$\begin{aligned} \Gamma_{21;1} &= \Gamma_{12;1} = -\Gamma_{11;2} = \frac{\partial a_{11}}{2\partial q_2}, \\ \Gamma_{31;1} &= \Gamma_{13;1} = -\Gamma_{11;3} = \frac{\partial a_{11}}{2\partial q_3}, \\ \Gamma_{41;1} &= \Gamma_{14;1} = -\Gamma_{11;4} = \frac{\partial a_{11}}{2\partial q_4}. \end{aligned} \tag{2.1.37}$$

Substituting the values into (2.1.36) we can write down the system of four equations for the motion of a resonant robot

$$\begin{aligned} &(a_{11}^0 + a_{11})\ddot{q}_1 + a_{13}^0\ddot{q}_3 + \frac{\partial a_{11}}{\partial q_2}\dot{q}_1\dot{q}_2 + \frac{\partial a_{11}}{\partial q_3}\dot{q}_1\dot{q}_3 + \frac{\partial a_{11}}{\partial q_4}\dot{q}_1\dot{q}_4 + b_1\dot{q}_1 = \\ &= D_1\rho_{02}/\rho_{01} - C_1 q_1, \\ &a_{22}^0\ddot{q}_2 - \frac{\partial a_{11}}{2\partial q_2}\dot{q}_1^2 + b_2\dot{q}_2 = \\ &= D_2/\rho_{11} - (m_2 + m_3 + m_4 + m_{21} + m_{31} + m_{32})g - C_2(q_2 - A_2), \\ &a_{33}^0\ddot{q}_3 + a_{13}^0\ddot{q}_1 - \frac{\partial a_{11}}{2\partial q_3}\dot{q}_1^2 + b_3\dot{q}_3 = D_3/\rho_{21} - C_3 q_3, \\ &a_{44}^0\ddot{q}_4 + \frac{\partial a_{11}}{2\partial q_4}\dot{q}_1^2 + b_4\dot{q}_4 = D_4 + M_y, \end{aligned} \tag{2.1.38}$$

where a_{ii}^0, a_{ik}^0 and a_{11} are given by (2.1.33), $M_y(q_4) = C_4 e_4(\rho_{31} + \rho_{32})\rho_{31}/\rho_{32} \times$ $\times\{1 - l_4[e_4^2 + (\rho_{31} + \rho_{32})^2 + 2e_4(\rho_{31} + \rho_{32})\cos q_4\rho_{31}/\rho_{32}]^{-1/2}\}\sin q_4\rho_{31}/\rho_{32}$.

2.1.8. From the form of Lagrange's equation (2.1.38) it is possible to identify the existing dynamic links between the robot's degrees of freedom. In the first equation, the variable a_{11} in the moment of inertia depends on the instantaneous value of the generalised coordinates q_2, q_3 and q_4. In addition, the torque relative to the rotation axis of the column appears to vary. The term $a_{13}^0\ddot{q}_3$ is caused by the tangential force due to inertia of the components, transferred by the third degree of freedom; $\frac{\partial a_{11}}{\partial q_2}\dot{q}_1\dot{q}_2$, $\frac{\partial a_{11}}{\partial q_3}\dot{q}_1\dot{q}_3$ and $\frac{\partial a_{11}}{\partial q_4}\dot{q}_1\dot{q}_4$ are the effects of the Coriolis inertia forces due to motion along the first coordinate as the remaining drives move at the same time.

In the second equation the term $\frac{\partial a_{11}}{2\partial q_2}\dot{q}_1^2$ reflects the influence of the vertical constituent of the normal force of inertia on the lifting carriage drive as the column rotates.

The third equation reveals the influence of the tangential forces $\frac{\partial a_{11}}{2\partial q_3}\dot{q}_1^2$, $a_{13}^0\ddot{q}_1$, which are the horizontal components of the normal force of inertia during rotation about the first axis of motion, on the drive of the horizontal arm. The fourth equation shows the influence on the rotational drive of the gripper, of a torque due to the normal inertia force as the drive of the first degree of freedom operates simultaneously.

The dynamic interactions between the drives considerably complicate robot control and bring about the need to oversupply driver power requirements.

The trend for robots to have higher operating speed causes them to develop considerable inertia forces. Hence in (2.1.38) the terms due to interactions between the motions can reach the same order as the basic inertia forces. As a result it becomes impossible to balance the inertia forces completely by using elastic spring forces or to relieve the drive load completely.

The influence of interacting movements in the MARS robot can be rationalised by the choice of parameters for its structural elements. Thus, for the elements *4*, *11*, *31* and *32* (Fig. 2.1.1), which are rotating bodies, it is not difficult to ensure that their moments of inertia relative to the perpendicular axes x and y are equal (the z-axis is parallel to the rotational axis of the elements): , $J_{x4}=J_{y4}$, $J_{x11}=J_{y11}$, $J_{x31}=J_{y31}$, $J_{x32}=J_{y32}$. In this case the value a_{11} depends only on the coordinate q_3 and the system of equations (2.1.38) becomes

$$\begin{aligned}
&(a_{11}^0+a_{11})\ddot{q}_1+a_{13}^0\ddot{q}_3+\frac{\partial a_{11}}{\partial q_2}\dot{q}_1\dot{q}_2+\frac{\partial a_{11}}{\partial q_3}\dot{q}_1\dot{q}_3+b_1\dot{q}_1=\\
&\quad=D_1\rho_{02}/\rho_{01}-C_1q_1,\\
&a_{22}^0\ddot{q}_2+b_2\dot{q}_2=\\
&\quad=D_2/\rho_{11}-(m_2+m_3+m_4+m_{21}+m_{31}+m_{32})g-C_2(q_2-A_2),\\
&a_{33}^0\ddot{q}_3+a_{13}^0\ddot{q}_1-\frac{\partial a_{11}}{2\partial q_3}\dot{q}_1^2+b_3\dot{q}_3=D_3/\rho_{21}-C_3q_3,\\
&a_{44}^0\ddot{q}_4+b_4\dot{q}_4=D_4+M_y,
\end{aligned}\qquad(2.1.39)$$

Here interrelationships arise only between the first and third equations. Their connection can be weakened by the elimination of torques $a_{13}^0\ddot{q}_1$ and $a_{13}^0\ddot{q}_3$, due to the tangential forces of inertia from the components. Providing MARS with a second arm, identical to the first, arranged as a mirror image of the first relative to the rotational axis of the column ensures that coefficient $a_{13}^0 = 0$. However, as a result the robot then has six degrees of freedom, the system of equations (2.1.39) increases by two equations (similar to the third and fourth) and, in addition, the inertia masses $(a_{11}^0 + a_{11})$ and a_{33}^0 are considerably increased. This also leads to considerable increases in the loads and a reduction in the speed of operation of the first axis drive.

There are two options to eliminate completely the dynamic interactions of motion between the degrees of freedom of the robot. The first option, which is achieved by cyclic resonant robots, consists of limiting the cycle of motion of the robot successively through all degrees of freedom. As a result a_{11} will remain constant during the motion process, the system of equations (2.1.38) will reduce to separate unrelated equations and will take the following form:

$$
\begin{aligned}
&(a_{11}^0 + a_{11})\ddot{q}_1 + b_1\dot{q}_1 = D_1\rho_{02}/\rho_{01} - C_1 q_1,\\
&a_{22}^0\ddot{q}_2 + b_2\dot{q}_2 =\\
&\qquad = D_2/\rho_{11} - (m_2 + m_3 + m_4 + m_{21} + m_{31} + m_{32})g - C_2(q_2 - A_2), \qquad (2.1.40)\\
&a_{33}^0\ddot{q}_3 + b_3\dot{q}_3 = D_3/\rho_{21} - C_3 q_3,\\
&a_{44}^0\ddot{q}_4 + b_4\dot{q}_4 = D_4 + M_y.
\end{aligned}
$$

The second option consists of making kinematic schemes for the robot with drives based on foundations. Such a method for constructing the kinematic arrangement is used, principally, in slave manipulators but complicates strongly the design and dynamics of automatic manipulators (robots).

It must be mentioned that the effects of motion interaction can even be beneficial in increasing the speed of operation of resonant robots. For example, in the case of simultaneous rotation of the column and displacement of the arm, driver *D1* can add energy to the third drive. The algorithm for interaction between the drives is similar here to that described in §1.5 shown in Fig. 1.5.15 (*Patent USSR 1171306, IPC B25J9/00*).

Since the majority of resonant robots function cyclically with successive motions through all degrees of freedom, we shall look in greater detail at the dynamics of a separate resonant drive.

§2.2 Resonant systems with ideal drives

Cyclic manipulator systems (tippers, shifters, etc) are widely used in industry because of their simplicity and reliability. Such devices usually operate at high speeds because powerful drivers are used. The substitution of resonant machines for the traditional variety makes even higher speeds of operation possible and reduces energy consumption during operation. We shall estimate the energy efficiency, speed of operation and loading in a resonant robot by comparison with the usual cyclic manipulator systems [5]. We shall assume that in both cases ideal drives are used.

2.2.1. We shall consider the *cyclic manipulator system* (Fig. 2.2.1) for a transfer device to move a workpiece of mass m_1 from zone *A* to zone *B*. For simplicity we shall assume that the arm of mass m_0 is displaced by the action of a constant value force F, produced by an idealised drive. The system operates in

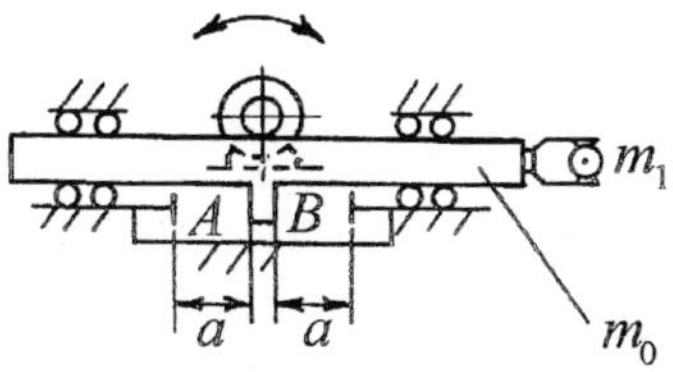

Fig. 2.2.1

the following manner: From zone *A*, the arm with workpiece moves to the stop *B* under the action of force F_1, is latched and pauses there for time τ_1 to unload the workpiece. After this, under force F_1 the arm returns to stop *A* where it pauses for the time τ_2 in order to load a new workpiece. The complete cycle is then repeated.

The motion of the arm of mass m_0 with workpiece of mass m_1 from stop *A* to stop *B* can be described by an equation of the form

$$m\ddot{x} + b\dot{x} + N = F_1, \tag{2.2.1}$$

where $m = m_0 + m_1$, x is the displacement of the arm, b is the viscous damping coefficient, N is the dry friction force, F_1 is the force causing motion.

After integrating (2.2.1) using the initial conditions $x(0) = -a,\ \dot{x}(0) = 0$ we find

$$\begin{aligned} \dot{x} &= (F_1 - N)[1 - \exp(-bt/m)]/b, \\ x &= -a + (F_1 - N)\{t - m[1 - \exp(-bt/m)]/b\}. \end{aligned} \tag{2.2.2}$$

The net time t_f taken for the displacement (neglecting dwell time) is found by calculation from the second equation of (2.2.2) when $x(t_f) = a$.

Mechanical systems of the type considered are of non-resonant form since the external force F_1 compensates not only energy losses in the system but also

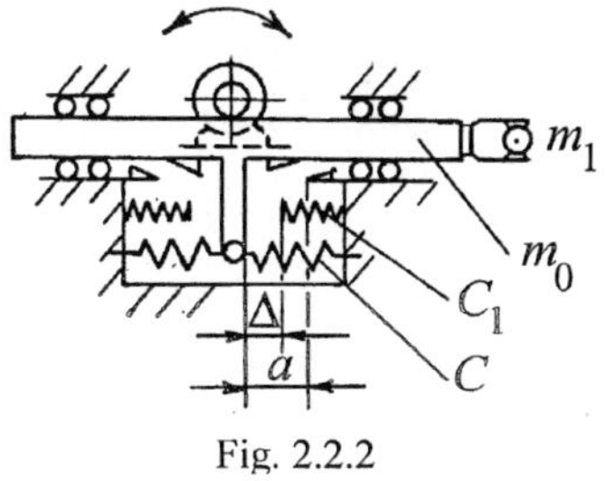

Fig. 2.2.2

inertia forces.

2.2.2. In order to construct a *resonant manipulator system* we introduce a central spring C. To stop the arm at the end of its travel we introduce elastic energy stores, characterised by their elastic coefficient C_1 (Fig. 2.2.2). The energy stores have latches that prevent the arm from returning after stopping. The resonant arm operates as follows: when the arm reaches zero velocity at the end of displacement, its kinetic energy is transferred into potential energy in one of the elastic energy stores; after a certain time has passed (τ_1 or τ_2) the latch is released and the stored energy returns the arm. If the elastic coefficient C_1 is chosen sufficiently large, then the time for transfer of energy between the store and arm is negligibly small by comparison with the time for motion from zone A to zone B and back.

Taking into account that the pause time (τ_1 or τ_2) does not affect the dynamic properties of the system examined, it can be considered, for modelling purposes, as a vibro-impact system [17], chosen with variable mass and double-sided limited action. The motion of the resonant system can then be described by the differential equations of the vibro-impact system modelled with additional pause times.

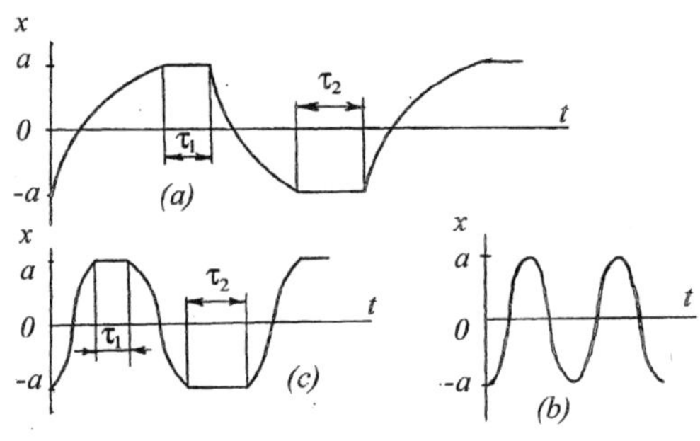

Fig. 2.2.3

Qualitative diagrams for non-resonant, resonant and the modelled vibro-impact systems are shown in Fig. 2.2.3 (*a*, *b* and *c* respectively).

The equation of motion of the vibro-impact system modelled is written in the form

$$[m_0 + m_1(1 + \operatorname{sgn}\dot{x}/2)]\ddot{x} + b\dot{x} + C(x - x_0) + \\ +(N - F)\operatorname{sgn}\dot{x} + H = 0, \tag{2.2.3}$$

where $H = H(x) = C_1(x+\Delta)\eta(-x-\Delta) + C_1(x-\Delta)\eta(x-\Delta)$ is the force characteristic of impact interaction; $\pm\Delta$ is the coordinate of the undeformed energy store; $\eta(y) = \begin{cases} 1, \text{ when } y > 0, \\ 0, \text{ when } y < 0, \end{cases}$ is the Heaviside function.

To determine the amplitude and frequency of vibration of the systems, we will use the harmonic linearisation method. We will search for a solution of the type

$$x \approx m_x + a_x \sin\omega t\,, \tag{2.2.4}$$

where m_x is a constant term; a_x is the amplitude of the fundamental; ω is the frequency of vibration.

We shall substitute for the nonlinear functions in (2.2.3) their harmonically linearised approximations

$$0.5m\,\mathrm{sgn}\,\dot{x} \approx \gamma\dot{x},\ \ H = H(x) \approx m_F + k(x - m_x),\ \ F\,\mathrm{sgn}\,\dot{x} \approx a\dot{x},$$
$$N\,\mathrm{sgn}\,\dot{x} \approx \beta_1\dot{x}. \tag{2.2.5}$$

From equation (2.2.3), taking account of (2.2.4) and (2.2.5), using the harmonic balance method we obtain the equations to find the parameters of the vibration fundamental

$$C + k - (m_0 + m_1/2)\omega^2 = 0,$$
$$b + \beta_1 = a,\ \ C(m_x + x_0) + m_F = 0. \tag{2.2.6}$$

We will calculate the linearisation coefficients for the nonlinear functions (2.2.5). Using a standard procedure [27] to find the harmonic linearisation coefficients for the dry friction force N and the excitation force F gives

$$a = 4F/(\pi a_x\omega),\ \ \beta_1 = 4N/(\pi a_x\omega). \tag{2.2.7}$$

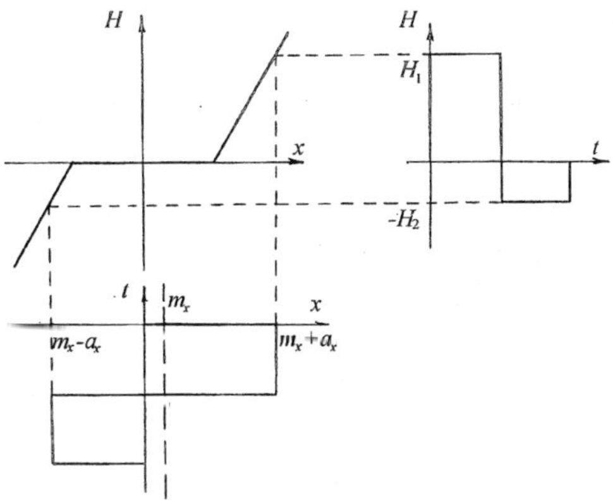

Fig. 2.2.4

To simplify the relationships found during linearisation of the elastic components of the forces of interaction between the arm and the elastic energy stores we will calculate (Fig. 2.2.4) the major coefficients of the linearisation function $H(x)$ in stepwise motion ("square sine wave")

$$m_F = \frac{1}{2\pi}\int_0^{2\pi} H d\psi = \frac{1}{2\pi}\left[\int_0^{\pi} H_1 d\psi - \int_{\pi}^{2\pi} H_2 d\psi\right] = C_1 m_x,$$
$$k = \frac{1}{\pi a_x}\int_0^{2\pi} H\sin\psi d\psi = \frac{1}{\pi a_x}\left[\int_0^{\pi} H_1\sin\psi d\psi - \int_{\pi}^{2\pi} H_2\sin\psi d\psi\right] = \tag{2.2.8}$$
$$= 4C_1(a_x - \Delta)/(\pi a_x),$$

where $H_1 = C_1(m_x + a_x - \Delta)$; $H_2 = C_1(-m_x + a_x - \Delta)$.

From the system of equations (2.2.6) using equations (2.2.7) and (2.2.8) we find the amplitude and frequency of the vibration fundamental and the dynamic offset m_x

$$a_x = \frac{2C_1\Delta}{4C_1 + \pi C} + \left[\left(\frac{2C_1\Delta}{4C_1 + \pi C}\right)^2 + \frac{16(m_0 + m_1/2)(F-N)^2}{\pi b^2(4C_1 + \pi C)}\right]^{1/2}, \tag{2.2.9}$$

$$m_x = Cx_0/(C_1 + C), \; \omega = 4(F-N)/(\pi a_x b).$$

Taking the last expression of the system (2.2.9) we find the actual time of movement in the resonant system, is equal to the half-period of vibration of the vibro-impact system modelled

$$T_p = \frac{\pi^2 a_x b}{4(F-N)}. \tag{2.2.10}$$

2.2.3 We shall now compare a non-resonant system with the resonant system equipped with an ideal drive.

From (2.2.9) it follows that the unique conditions for the existence of self-sustaining vibration is to ensure that the excitation force dominates the dry friction force. For a non-resonant system, as can be seen from (2.2.2), the condition $F > N$ must be satisfied as well.

To compare the energy and force factors of the two systems we shall assume for example that $t_f = T_p = m/b,\ a_x = a$. Then from (2.2.10) and (2.2.2) we find

$$F_1 - N = 2eab/t_f, \;\; F - N = \pi^2 ab/(4t_f), \tag{2.2.11}$$

where e is the base of natural logarithms. The ratio $(F_1 - N)/(F - N) = 8e/\pi^2 \approx 2.2$ shows by how much the non-resonant system is overloaded. The energy expenditure on transportation for both systems is calculated from the formula

$$(A_1 - A_N)/(A - A_N) = (F_1 - N)\int_{-a}^{a} dx/(F-N)\int_{-a}^{a} dx \approx 2.2, \tag{2.2.12}$$

where F_1 and F are found from (2.2.11). (2.2.12) shows the amount by which the power consumption of the cyclic drive exceeds that of the resonant system.

To compare the operating speeds we put $F_1 = F = N + 2eab/t_f$. In this case

$$t_f / T_p = 8e/\pi^2 \approx 2.2. \tag{2.2.13}$$

In this way, we shown that by replacing a non-resonant cyclic system with a resonant system, it is possible:

- to unload the drive by a large factor and to reduce its energy consumption while preserving its operating speed ($t_f = T_p$), by suitable choice of the elasticities C_1 and C in (2.2.9);
- to increase the system's operating speed considerably by maintaining the initial drive ($F_1 = F$) and the level of energy consumption.

The results of this comparison reflect the essence of a resonant system, in which the operating motion is not very different from free vibration of the mechanical parts of the drive while control is aimed at compensating the dissipative losses in the drive. Friction forces in the drive are low by comparison with the inertia forces and spring elasticity. Comparatively low control effects are required to generate motion. For this reason resonant systems are called "weak controllable systems" [69].

We must not forget that this section examined the simplest non-resonant cyclic system (Fig. 2.2.1), which only has no arrangement to slow it down as it approaches the stops. As a result the manipulator arm will impact against the stops. In actual cyclic systems damping becomes established at the end of travel (§1.3) or opposing drivers are supplied to produce the same uniform approach to the stops. In these cases, the energy factor and operating speed are considerably worse than for the cyclic system considered.

§2.3 Electromechanical drive for resonant systems

As was explained in the preceding section, in resonant robotic systems with idealised drive, the forces produced by the drive are used only to compensate for the energy losses in the system and to produce useful work; inertia forces in such systems are balanced by elastic elements. By ensuring high mechanical Q-factor in the vibrating resonant system, the speed of operation of a cyclic robot can be considerably increased or, at the required operating speed, the required drive power can be reduced considerably.

2.3.1. As an example of a module (Fig. 2.2.1, where $C_1 = 0$) for the extending arm of the MARS robot (§2.1) we shall consider a resonant system with electric drive [4, 7]. A description of the operation of the module has been given earlier in §§2.1 and 2.2.

Since pausing at the extreme positions does not change the dynamic conditions of the system, calculation will be based on the basic study of the self-sustaining vibration system, causing continuous motion and described by the equation

$$m\ddot{x} + b\dot{x} + Cx + N\,\mathrm{sgn}\,\dot{x} = D/r\,, \tag{2.3.1}$$

where $m = m_0 + m_1(1+\mathrm{sgn}\,\dot{x})/2$; m_0 is the mass of the arm and the reduced mass of the driver rotor; m_1 is the mass of the load being moved; x is the arm coordinate, measured from its position when the spring with elasticity C is in equilibrium, N is the absolute value of the dry friction force; r is the radius of the operating wheel; D is the torque provided by an independently excited direct-current electric motor.

From (1.3.4), (1.3.5) and (1.3.7) we find

$$D = ki = k(u-E)/R\,, \tag{2.3.2}$$

where k is a construction parameter of the electric motor; u and i are the driving voltage and current in the driver armature; $E = k\dot{x}/r$ is the emf induced in the armature.

When solving the system of equations (2.3.1) and (2.3.2) which describe the dynamic state of the resonant module with electric motor, we shall consider forward ($\dot{x} > 0$) and reverse ($\dot{x} < 0$) motion of the robot arm separately.

Forward motion of the robot arm is described by the linear differential equation

$$\ddot{x} + 2n\dot{x} + \omega^2 x = F\,, \tag{2.3.3}$$

where $n=(b+k^2/Rr)/2(m_0+m_1)$ is the damping coefficient, $\omega^2=C/(m_0+m_1)$ is the square of the natural frequency of the system taking account of the mass of the load, $F=[ku/(Rr)-N]/(m_0+m_1)$.

Taking into account that forward motion starts from rest when the coordinate of the moving component is found from the position of contact with the left-hand latch, we have

$$x(0)=-a_1,\quad \dot{x}(0)=0\,. \tag{2.3.4}$$

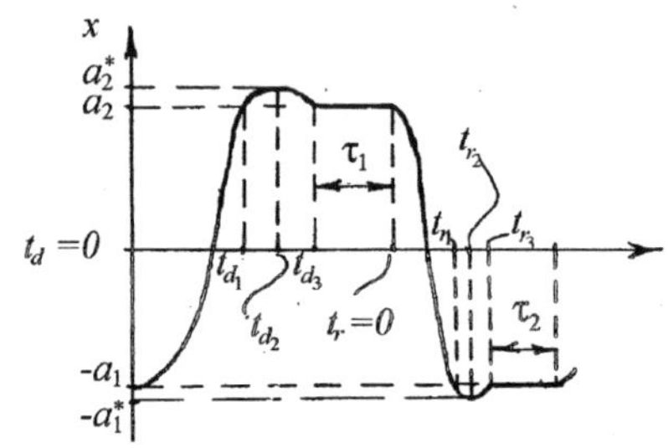

Fig. 2.3.1

Bearing in mind that the natural condition for resonant of the system with considerable oscillation is the regime $n<\omega$ and taking account of the initial condition (2.3.4) we can write the solution of equation (2.3.3) in the form

$$x=Be^{-nt}\sin(\Omega t+\psi)+F/\omega^2\,, \tag{2.3.5}$$

where $B=-a_1\omega/\Omega-F/(\omega\Omega)$, $\psi=\arctan\Omega/n$, $\Omega^2=\omega^2-n^2$.
Similarly we can write the equation for reverse motion (initial conditions $x(0)=+a_2$, $\dot{x}(0)=0$). Using these equations and taking account of the pause time produces the working diagram for the module (Fig. 2.3.1).

2.3.2. During the study of the equations of motion of the robot arm it was assumed that before motion started it is in one of the end positions and the spring is fully charged with potential energy. The module will operate normally only when these conditions are satisfied. If, as the result of malfunction, (for example if the circuit voltage fails) the arm remains in the central position then simply re-applying this voltage will not ensure its return to the initial position, because the use of a low powered drive will not overcome the spring force. For this we must use a drive control (Chapter 3) to swing the robot arm relative to its static equilibrium position and to ensure smooth return of the arm onto one of the latches.

Malfunction prevention is related to finding *stable regimes of normal operation.* For such a study account is taken that each cycle of motion begins with identical initial conditions found from the position of the latches and does not depend on the excitation affecting the previous cycle. Consequently, the operating regime will be stable in the case when the robot arm, during forward motion, deviates from the equilibrium position by an amount greater than a_2, and during reverse motion by an amount not less than a_1. In fact such conditions of stability for the normal operating regime coincide with the conditions for this regime to exist. As a measure of stability the characteristic excitation band during which the operation of the module survives can be taken as the parameter $\eta = E_{kin} / E_{pot}$ where E_{kin} and E_{pot} are the kinetic and potential energies of the arm as it approaches the defined latch at the end position. When $\eta = 0$ the motion is unstable, and when $\eta > 0$, it is stable, from which we can deduce that the larger the value of η then the greater is the stability margin. The physical sense of the stability conditions is that energy transferred to the moving component from the driver for one complete cycle must be greater than that dissipated.

In the cyclic diagram of the robot module (Fig. 2.3.1), the sections $t_d = 0 - t_{d2}$ and $t_r = 0 - t_{r2}$ describe the forward and reverse motion while the sections $t_{d2} - t_{d3}$ and $t_{r2} - t_{r3}$ represent the same motion but when the applied force N has the opposite sign. Sections τ_1 and τ_2 correspond to the pause time and sections $t_{d1} - t_{d3}$ and $t_{r1} - t_{r3}$ the time needed to ensure the stability margin. Thus if the stability margin is small the robot module can malfunction while an unnecessarily large stability margin causes not only excessive energy use but also increases the overall cycle time (and correspondingly reduces the productivity).

2.3.3. In the design of resonant modules the problem arises of rational choice of driver and, corresponding to this, the optimal drive transmission ratio (radius r of the operating wheel). Using the operating requirements of the module when the drive is only required to compensate dissipative losses, we will take radius r_* as the optimum radius of the operating wheel (the base circle radius in the case of gear transmission), which ensures the transmission of maximum energy to the moving component.

To obtain an approximate solution we will assume that motion in both forward and reverse directions is close to simple harmonic motion with frequencies equal to the natural frequencies during an appropriate motion. We will also assume that the end latches are arranged symmetrically relative to the central position of the spring, i.e. $a_1 = a_2 = a$. In addition, we will assume that the mass m_1 of the load carried is small compared to the mass m_0 of the arm and, consequently, differences between the forward and reverse motions can be neglected. Hence, to choose driver and operating wheel we will consider a single forward stroke of the robot arm. The energy transmitted to the robot arm from the driver for a forward stroke is

$$A = \int_{-a}^{a} \frac{D}{r} dx = \int_{0}^{\pi/\omega} \frac{D\dot{x}}{r} dt = \frac{ka}{Rr}(2u - \pi ka\omega / 2r), \tag{2.3.6}$$

where $D = k(u - k\dot{x}/r)/R, \quad \dot{x} = a\omega \sin \omega t$.

Then the optimal radius of the operating wheel is found from the equation

$$r_* = \max_r A = \pi ka\omega / 2u. \tag{2.3.7}$$

Using (2.3.2), we can express r_* in terms of the data supplied for an electric motor - D_n and i_a the nominal torque and armature current respectively

$$r_* = \pi a\omega D_n / 2ui_a \,. \tag{2.3.8}$$

It can be seen from these expressions that the torque acting on the robot arm from the drive remains negative during some period of time when $r < r_*$ (Fig. 2.3.2). During this time the driver operates as a generator [37], recovering

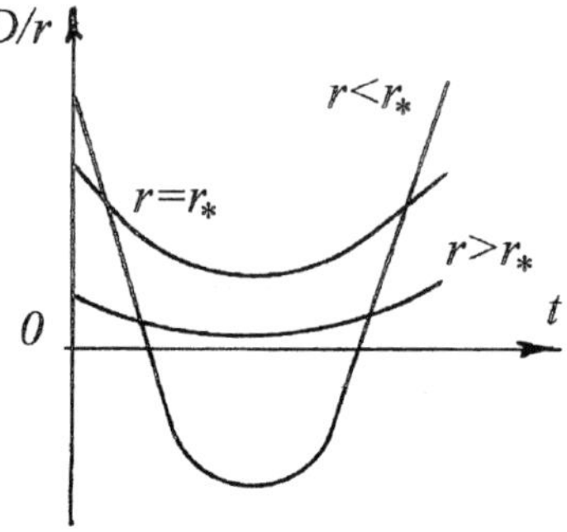

Fig. 2.3.2

supplied energy as the robot arm accelerates the driver to a speed greater than its no-load speed by means of the spring. When $r > r_*$ the force applied to the robot arm remains small and is not time dependent in practice.

The choice of electric motor drive parameters will be made from the expression $A_* = A_0 + \eta Ca^2/2$. This takes into account that the energy A_* received by the robot arm from the driver during each displacement must exceed the energy dissipated in the system by some amount $\eta Ca^2/2$, to ensure the stability margin for the mode of operation. The value of A_* is calculated from (2.3.6) when $r = r_*$

$$A_* = 2u^2 / \pi R\omega \,. \tag{2.3.9}$$

The work due to the dissipative forces is found from the formula

$$A_0 = \int_0^{\pi/\omega} (N\dot{x} + b\dot{x}^2)dt = 2Na + \pi a^2 b\omega/2 \,. \tag{2.3.10}$$

As a result we obtain the expression

$$u^2 / R = \pi a\omega[N + a(\pi b\omega + \eta C)/4] \,, \tag{2.3.11}$$

which makes it possible to select the two main parameters of the electric motor - the supplied voltage u and the active resistance of the armature R - from the known structural parameters of the mechanical parts of the resonant system. After this and based on the supplied data for the chosen electric motor we can find the corresponding radius r_* from (2.3.7).

Taking into account that, during the design stage, various structural parameters of the system (N, b, C and others) can only be measured approximately, then it is reasonable to consider an increase of 10 - 15% in the supplied voltage when choosing an electric motor from catalogued data.

Chapter 3. Optimal control of resonant systems

§3.1 Statement of the optimisation problem. Efficiency criteria

The general statement of the problem of dynamic analysis and synthesis of resonant robotic systems involves a search for the optimal structure and control of the robot with respect to certain efficiency criteria. Resolving the optimisation problem makes it possible to study the limiting values of resonant robot dynamics, to compare different technical solutions and to make a rational choice for the arrangement and control of a robot. As a result it is possible to avoid unjustified high expenditure on the production of technically complex cyclic robots, when dynamic qualities are only improved by trivial amounts.

A test to compare the efficiency of an ideal resonant drive with a traditional cyclic drive was introduced in §2.2. It was shown that a resonant drive is superior to the traditional one both in energy consumption and operating speed by several hundred per cent. To control this a constant load from an ideal actuator was considered. However the question of efficiency of such control (or any other of an assortment of possibilities) compared to the optimal was left unanswered.

In what follows primary attention will be given not to the whole robot but to the separate resonant drive since a resonant robot operates in a cyclic manner (operating motion is produced consecutively along all degrees of freedom).

Synthesis of an optimal drive for a resonant robotic system is aimed at obtaining the optimal drive structure and optimal control of this drive.

We introduce a standard method of formulating optimal the control problem [84]:

It is required to determine the vector-function of variable states - $\varphi(t) \in \mathbf{R}^n$ and the control actions $u(t) \in \mathbf{R}^m$ when $t \in [t_0,\ t_f]$, providing the minimum of the criterion functional

$$\mathfrak{J} = \mathfrak{J}(\varphi, u, t) \tag{3.1.1}$$

under the differential conditions

$$\dot{\varphi} = f(\varphi, u, t), \tag{3.1.2}$$

limited along the trajectory

$$(\varphi, u, t) \in \mathfrak{U} \tag{3.1.3}$$

and with boundary conditions

$$\mathfrak{G}\left(t_0, \varphi(t_0);\ t_f, \varphi(t_f)\right) = 0\,. \tag{3.1.4}$$

We assume that f is a continuous $n-$vector function, continuously differentiable over all variables; the set $\mathfrak{u}$ represents a domain of space $\mathbf{R}^n \times \mathbf{R}^m \times \mathbf{R}^l$; $\mathfrak{G}$ is an l-dimensional vector ($l \leq 2n+1$). The control u can be an arbitrary piecewise-continuous function in the set $\mathbf{R}^m$.

In this chapter we will construct optimal controls for two general types of resonant drive (cyclic and stepped), to ensure the extreme values of the functional (3.1.1) and taking account of the peculiarities of resonant robots.

As an example of a cyclic device, we shall consider a sample resonant module (Fig. 3.1.1) in which the robot swinging arm *1*, is kinematically linked to the direct

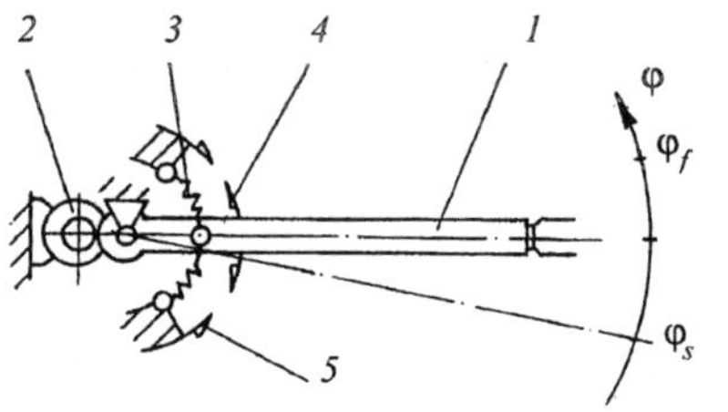

Fig. 3.1.1

current electric motor *2* through gears with transmission ratio *Z*. The mechanical accumulator consists of linear springs *3* with overall stiffness C, connecting arm *1* to the body. Latches *4* and *5* hold the robot arm at the end points in positions φ_s and φ_f.

The multi-position stepped module (Fig. 3.1.2) with resonant drive consists of

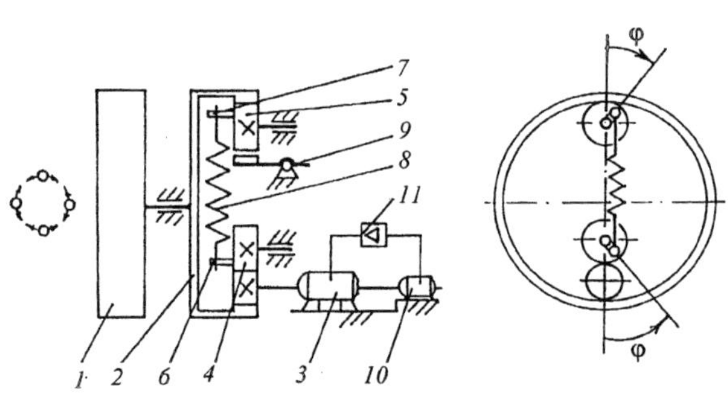

Fig. 3.1.2

a driven member, disc *1*, which is connected through gear *2*, having internal teeth, to the electric motor *3*, and through two identical gears *4* and *5* to the cranks *6* and *7*, whose ends are interconnected by the linear tension spring *8* (*Patent USSR 1544550, IPC B25J9/00*). The cranks are held in unstable equilibrium by latch *9*, with the spring elongated to its maximum. Each rotation of the cranks corresponds to a single step of the disc.

Descriptions of the operation of these cyclic and stepped modules are given in §1.5.

Assuming that the stiffness of the module transmission is considerably greater than the equivalent stiffness of the accumulator spring, we will use a kinetostatic model for the drive (§1.3). Furthermore, based on the fact that the speed of electromagnetic processes in a direct current electric motor with independent excitation is substantially greater than the speed of motion and rate of change of loads in the transmission, we shall use static characteristics (1.3.7) to represent the actuator. The accuracy of the last assumption can be proved from a study of the dynamic system using frequency methods. In particular, it was shown in [60] that it is not necessary to consider jointly the combination of strongly interconnected sub-systems in the case where the velocities of operating processes occurring in the subsystems are strongly different.

Technology requirements (determined by the primary industrial processes for which a robot is used) and service conditions (such as operating temperature, water protection, fire-proofing, floor- or wall-mounting, etc.) together with the specific properties of the robotic system (economy, operating speed, response, accuracy, load capacity, etc.) predetermine the selection of a quality criteria set [35, 48, 62, 91, 103].

A resonant manipulator system (Figs. 3.1.1 and 3.1.2) is intended for the transport of a workpiece from an initial position φ_s to a final position φ_f over a period of time t_f. The velocity of the object at the beginning and the end of this period must be zero. During this time, the drive, its control and character of motion must satisfy a multitude of requirements, each of which can be formulated as equations in the form of minima of some target function (functional) or in the form of an inequality. In the general case the minimised functional (3.1.1) is represented in the form [84]

$$\mathfrak{J} = \mathfrak{G}\left[\varphi(t_0),\ \varphi(t_f)\right] + \int_{t_0}^{t_f} \mathfrak{Q}\left[\varphi(t), u(t), t\right] dt\ , \tag{3.1.5}$$

where the symbol $\mathfrak{G}$ is a scalar function of the boundary states of the system; $\mathfrak{Q}$ is a scalar function, which depends on the current time t, phase coordinates φ and controls u. We shall define the quality parameters (criterion functionals) of the resonant robotic systems and their controls [19-23, 62, 91, 96, 97, 103]:

a) The *operating speed of a resonant drive* is characterised by the time t_f for movement from initial position φ_s to final position φ_f

$$\mathfrak{J}_1 = \int_0^{t_f} dt = t_f\ . \tag{3.1.6}$$

b) The *total energy consumption of the direct current motor with independent excitation.* We can write the equation for electrical equilibrium of the control circuit of the actuator armature components in the form [37]

$$u = E + Ri\ . \tag{3.1.7}$$

Equation (3.1.7) is a simplified form of (1.3.3) when $L << R$.

The equation for the power balance is found by multiplying the left and right hand sides of (3.1.7) by the current i

$$ui = Ei + Ri^2\ . \tag{3.1.8}$$

Equation (3.1.8) indicates that the motor operates as a driver. As a result, the power consumed (ui) is partly lost in the form of heat due to resistance in the circuit components (Ri^2) and partly converted into electromagnetic power (Ei). The energy consumption of the actuator over time t_f can be put in the form

$$\Im_2 = \int_0^{t_f} uidt = \int_0^{t_f} (Ei + Ri^2)dt . \qquad (3.1.9)$$

Or, by using (1.3.7) we find

$$\Im_2 = \int_0^{t_f} (Rk^{-2}Q^2 + Q\dot{\varphi})dt . \qquad (3.1.10)$$

It is obvious that the composite functional $\Im_2$ is a combination of the functionals

$$\Im_3 = \int_0^{t_f} Rk^{-2}Q^2 dt , \qquad (3.1.11)$$

which represents the motor *heat generation*, and

$$\Im_4 = \int_0^{t_f} Q\dot{\varphi}dt , \qquad (3.1.12)$$

which represents *mechanical work produced by the electric motor*.

c) The *operating speed of a resonant drive* can also be evaluated indirectly from the area under its phase curve

$$\Im_5 = \int_{\varphi_s}^{\varphi_f} \dot{\varphi}d\varphi = \int_0^{t_f} \dot{\varphi}^2 dt . \qquad (3.1.13)$$

The resonant system is cyclic - either oscillatory (curve *1* in Fig. 3.1.3) or stepped - rotational (curves *2* and *3* in Fig. 3.1.3). From physical reasoning it is obvious

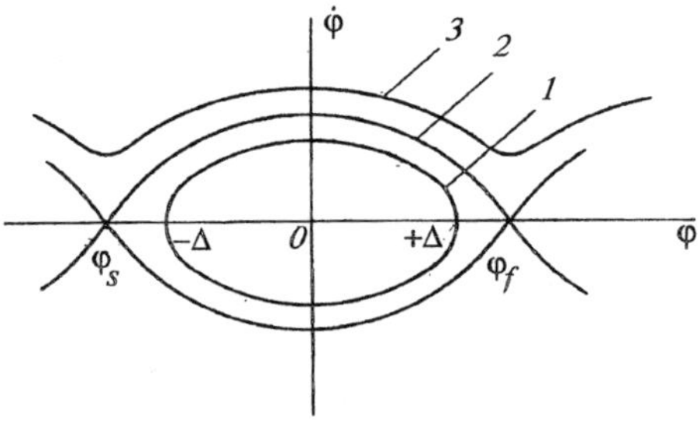

Fig. 3.1.3

that by ensuring maximum velocity $\dot{\varphi}$ on all paths (φ_s to φ_f) of motion for the oscillating system, we will ensure minimum displacement time (maximum operating speed and area under the phase curve). From the maximum principle [86] (which we will use later to define optimal controls) it follows that the control sought will ensure that the term under the integration sign in the functional is a maximum at almost every point φ in the trajectory. Hence the desired optimal control results in a system with maximum speed at almost every instant.

d) *Cost of reaching the end points*. The stability of operation of a resonant drive for a manipulator system can be characterised by the ratio of the kinetic energy E_{kin} to the potential energy E_{pot} of the driven member as it approaches the final latched position [7]. The operational conditions for a cyclic resonant drive are such that this ratio can vary somewhat. Thus in the cyclic manipulator module (Fig. 3.1.1) the kinetic energy for the advance of the arm to the spring latch must, on the one hand, be enough to open the latch, but, on the other hand, must not provide significant overtravel of the arm. Therefore, to describe stability of operations it is convenient to consider the ratio of the deviation of the actual value of kinetic energy $0.5J\dot{\varphi}^2(T)$ from the required value of $0.5J\dot{\varphi}_T^2$ in the latched position, to the value of potential energy E_{pot} in the mechanical accumulator in the latched position. Bearing in mind that the values of J and E_{pot} are fixed, then the characteristic for stable operation of the drive can be represented in the form of a functional

$$\Im_6 = \left|\dot{\varphi}^2(T) - \dot{\varphi}_T^2\right|. \tag{3.1.14}$$

e) The *cumulative torque of the driver momentum* is used as a criterion in a series of cases to estimate the energy consumption and size of the driver

$$\Im_7 = \int_0^{t_f} |Q|\, dt. \tag{3.1.15}$$

f) *Combining the criteria*. Minimising the functionals $\Im_1$ - $\Im_4, \Im_6, \Im_7$ and also $(-\Im_5)$ improves the quality of a resonant drive. However, a decrease in one of them can be accompanied by an increase in the others, for example, trying to ensure minimum motion time ($\Im_1$) will increase energy losses ($\Im_2$). For this reason, the functionals are combined, appearing as linear combinations of those above, with weighted coefficients depending on the chosen constraints. Thus, in the theory of optimal control, by analogy with classical problems of variational calculus [84], the following problems are distinguished:

- Bolza's problem, when $\mathfrak{G} \neq 0, \mathfrak{q} \neq 0$, from (3.1.5),
- Lagrange's problem, when $\mathfrak{G} = 0, \mathfrak{q} \neq 0$,
- Mayer's problem, when $\mathfrak{G} \neq 0, \mathfrak{q} = 0$.

Note that criteria of minimal-maximal type used when studying impact effects in vibration protection systems [53, 65] have not found wide applications in resonant robotic systems.

Apart from quality criteria, in problems of optimal synthesis of controlled resonant machines, constraints are imposed (3.1.3) on motion, transmission and driver parameters [84, 102, 103]. For example the use of serial electric motors and mechanical accumulators assumes constraints on the maximum allowable control voltage and current and also on the maximum possible operating forces in the accumulator. These conditions can be formulated in the form of various inequalities that must be satisfied by any controller and drive parameters.

The *edge conditions* (3.1.4) appear, as shown, in the form of the initial conditions

$$\varphi(t_0) = \varphi_s \tag{3.1.16}$$

and the boundary conditions

$$\mathfrak{G}_f\left(t_f, \varphi(t_f)\right) = 0 . \tag{3.1.17}$$

From this, the following problems can be distinguished:

- with fixed ends and given values of $\varphi(t_0)$ and $\varphi(t_f)$;
- with free right hand end, when $\varphi(t_f)$ is not defined.

Notice that in the majority of problems in optimal control of resonant systems, time t_f for the completion of motion is not fixed. This explains the main feature of resonant systems - to ensure that operating speed is determined by its natural frequency of vibration without the use of control. That is why time t_f is determined by the period of free vibrations or position $\varphi(t_f)$.

The general means for the solution of the problem of optimal control in resonant systems is to use Pontryagin's maximum principle [20-23, 29, 30, 34, 35, 38, 40, 54, 69, 86, 87, 90, 96, 97]. The aims of optimisation can also be achieved by using Bellman's dynamic programming method [13, 26, 41, 54, 86, 87, et al.] and methods based on the ideas of classical variational calculus [102, 103].

We will set out the wording of *Pontryagin's maximum principle,* defining the conditions for optimisation [86, 90].

In order to obtain permissible (within the imposed constraints) values for $u^*(t), \varphi^*(t), t_0^*, t_f^*$ as solutions to the general problem of optimal control, it is necessary that Lagrangian coefficients should exist

$$\lambda_i^*,\ i = 0, \ldots, l \quad \lambda_0^* \geq 0,\ \lambda^*(t),\ t \in [t_0^*,\ t_f^*],$$

which are not simultaneously equal to zero and also that when

$$\lambda = \lambda_i^*,\ i = 0, \ldots, l,\quad \lambda(t) = \lambda^*(t),\quad \varphi(t) = \varphi^*(t)$$

$$u(t) = u^*(t),\ t_0 = t_0^*,\ t_f = t_f^*$$

the following conditions are satisfied:

1)
$$\dot{\varphi} = f(\varphi, u, t) = \frac{\partial H}{\partial \lambda}, \tag{3.1.18}$$

$$\dot{\lambda} = -\frac{\partial H}{\partial \varphi} \tag{3.1.19}$$

for virtually all $t \in [t_0, t_f]$:

2)
$$\int_{t_0}^{t_f} \mathfrak{C}_i(\varphi, u, t)\,dt + \mathfrak{G}_i\left[\varphi(t_0), \varphi(t_f)\right] \begin{cases} = 0,\ i = 1, \ldots, l_1, \\ \leq 0,\ i = l_1 + 1, \ldots, l \end{cases} \tag{3.1.20}$$

3)
$$\lambda(t_0) = \frac{\partial}{\partial \varphi(t_0)} \sum_{i=0}^{l} \lambda_i \mathfrak{G}_i\left[\varphi(t_0), \varphi(t_f)\right], \tag{3.1.21}$$

$$\lambda(t_f) = -\frac{\partial}{\partial \varphi(t_f)} \sum_{i=0}^{l} \lambda_i \mathfrak{G}_i\left[\varphi(t_0), \varphi(t_f)\right], \tag{3.1.22}$$

4)
$$H\left[\varphi(t),u(t),\lambda(t),t\right]=0 \tag{3.1.23}$$
when $t=t_0$ and $t=t_f$;

5)
$$\lambda_i\left\{\int_{t_0}^{t_f}\mathfrak{G}_i(\varphi,u,t)dt+\mathfrak{G}_i\left[\varphi(t_0),\varphi(t_f)\right]\right\}=0 \tag{3.1.24}$$

$i=1,\dots,l,$ so that
$$\lambda_i\ge 0,\ i=l_1+1,\dots,l; \tag{3.1.25}$$
6) The function $H=\left[\varphi(t),u(t),\lambda(t),t\right]$ known as a Hamiltonian, considered as a function $u\in\mathfrak{u}$, has an absolute maximum
$$H\left[\varphi(t),u(t),\lambda(t),t\right]=\max_{u\in\mathfrak{u}}H\left[\varphi(t),u(t),\lambda(t),t\right] \tag{3.1.26}$$
for almost all $t\in[t_0,t_f]$, while, if $u(t)=u^*(t)$ in some interval of time, it takes the value of an internal point within the space $\mathfrak{u}$ so that in this interval
$$\frac{\partial H}{\partial u}=0\,. \tag{3.1.27}$$
If t_0 or t_f are fixed, then the corresponding condition (3.1.23) is not applied. If $\varphi(t_0)$ is fixed then condition (3.1.21) is not applied and if $\varphi(t_f)$ is fixed then condition (3.1.22) loses validity.

In the general case it is very rare to obtain an exact solution of the optimal control problem. However, for a resonant system which has weak control applied to its drive [34], it is very convenient to use asymptotic methods, matching the classical approach to optimal control with approximate methods from vibration theory [32-34, 36, 69, 79, 83, 86, 87]. Since control in resonant robotic system drives is created to take effect over finite time intervals, having an order of the period of oscillations then it is more efficient to use the method of successive approximations [69, 86, 87]. Using this, it is possible to obtain quasi-optimal control in the form of closed analytical solutions in a number of cases. Above all, results obtained by such methods can be used as initial approximations for numerical methods [39, 86].

§3.2 Synthesis of optimal cyclic drive

The synthesis of a manipulator drive considered as a controlled dynamic system, must be undertaken in the form of a joint search for the mutually conditioned structure and control of the drive. The optimal will be the dynamic system whose structure makes it possible to achieve strong effects by means of weak controls. In turn, the optimal control must be adequate for the chosen drive structure and must ensure extreme values for the efficiency criteria.

It is almost impossible to solve the problem of synthesis of an optimal drive for robots functioning with contour or positioning conditions. The way of programming the robots (in most cases by "teaching" the robot), is universally applied, and as a result, the uncertainty in defining a quality criteria set, causes difficulties in resolving the problem of simultaneously choosing optimal structure and control of the robot drive. For this reason, optimal controls are selected for the robot drives built on the basis of designer's preferences.

The unambiguous function of a cyclic manipulator drive, to generate forward and reverse (oscillating) transport motion between two fixed points φ_s and φ_f over a given time t_f, makes it possible to state the problem of synthesising the optimum drive in terms of operating speed and energy consumption [23].

From the dynamics of a cyclic drive with dominant inertia loads, rigid transmission and direct current motor with independent excitation, the motion of the driven component can be described by equation (1.3.2)

$$J\ddot{\varphi} + b\dot{\varphi} + N_1 \operatorname{sgn}\dot{\varphi} = ZQ\,, \tag{3.2.1}$$

with static characteristic (1.3.7) for the driver

$$Q = \frac{k}{R}u - \frac{Zk^2}{R}\dot{\varphi} \tag{3.2.2}$$

and the boundary conditions

$$\begin{aligned} &t = 0, \varphi = 0, \dot{\varphi} = 0, \\ &t = t_f, \varphi = \varphi_f, \dot{\varphi} = 0, \end{aligned} \tag{3.2.3}$$

where φ is the displacement of the driven component, J is the moment of inertia of the moving components of the drive transferred to the driven component, b is the coefficient of viscous friction, Z is the transmission ratio of the reduction gearbox, Q is the torque supplied by the driver, u is the control voltage, k is the motor parameter from (1.3.4) to (1.3.6), t_f is the cycle time. In what follows, we shall consider a cycle of motion when $\dot{\varphi}(t) \geq 0$.

During cyclic motion of a drive with boundary conditions (3.2.3) the instantaneous power $P_Q(t)$ can be separated into two components, namely the active $P_D(t)$ and reactive $P_M(t)$ components [12]. The active power from the motor generally consists of that required to replenish the energy irreversibly lost in the drive due to dissipation. The average value per cycle of the reactive power that is uselessly circulated in the transmission between the driver and the driven component is equal to zero. The value of the instantaneous powers can be described relative to the active $D_1(t)$ and reactive $M_1(t)$ torques (controls) in the form

$$\begin{gathered} P_Q(t) = P_D(t) + P_M(t), \\ \int_0^{t_f} P_Q dt = \int_0^{t_f} P_D dt, \qquad \int_0^{t_f} Q_1\dot{\varphi} dt = \int_0^{t_f} D_1\dot{\varphi} dt, \\ \int_0^{t_f} P_M dt = \int_0^{t_f} M_1\dot{\varphi} dt = \int_0^{t_f} M_1 d\varphi = 0. \end{gathered} \tag{3.2.4}$$

Since the reactive torque M_1 performs no mechanical work during the cycle of motion then, for economy in general energy expenditure in the drive, a mechanical energy store can be introduced in the drive structure, capable of forming the reactive torque. For example, the energy store can be implemented in the form of a flywheel with variable ratio drive to provide $M_1(t)$, or an elastic element with a force characteristic $M_1(\varphi)$. The use of flywheels as energy stores is preferable in many cases, because of their high energy capacity [46, 50, 52]. However, in the absence of reliable automatically controlled speed variators, potential energy stores are assumed in practice (§1.4) [51, 74, 80].

The active torque $D_1(t)$ must be generated by the driving electric motor (3.2.2).

We will choose as criteria to be minimised a combination of functionals characterised in the form (3.1.10) for the overall electrical energy consumption and the form (3.1.13) for the operating speed of the drive

$$\Im = RJ^2(Zk)^{-2}\int_0^{t_f}[D^2 + \alpha k^2 Z^2 (RJ)^{-1} D\dot{\varphi} - \mu\dot{\varphi}^2]dt. \qquad (3.2.5)$$

Here, $D = ZD_1J^{-1}$; t_f, the motion time, is not fixed; α and μ are non-negative weighting coefficients, various values of which can change the significance of the energy component and the term that characterises the operating speed of the drive in the functional. When $\alpha = \mu = 0$ the problem is reduced to the minimisation of heat losses in the electrical parts of the motor; when $\alpha = 1,\ \mu = 0$ the overall energy consumption of the motor is minimised; when $\alpha >> \mu$, the active power of the drive is minimised; when $\alpha << \mu$ the time of motion t_f is minimised.

Note that for the case when $\mu = 0$, the value of the functional with non-fixed motion time will not exceed the value of the functional when t_f is fixed. As a result, the value t_f^{opt} found can be used either to estimate the efficiency of the machine cycle time prescribed by the cyclogram, or in initial design.

From the form of the functional and the properties of the solution to the problem, it follows that the required optimal control D will be limited. For this reason we shall not impose any limitations on the form and value of the active control. It is convenient to impose a limit on the reactive control M of the form

$$|M| \le M_0 \qquad (3.2.6)$$

where $M = M_1J^{-1}$, M_0 is the maximum torque supplied by the energy store.

Thus, the initial dynamic problem (3.2.1) to (3.2.3) to find, matched in economy and operating speed, the structure and control of the cyclic drive can be formulated as the problem to find two optimal controls D_* and M_*, as in the following example.

For the drive, described by the equation

$$\ddot{\varphi} + 2n\dot{\varphi} + N = D + M, \qquad (3.2.7)$$

it is necessary to construct controls $D(t)$ and $M(t)$ that will change the driven component from the initial state

$$t = 0,\ \varphi = 0,\ \dot{\varphi} = 0,$$

to the final state

$$t = t_f,\ \varphi = \varphi_f,\ \dot{\varphi} = 0$$

provide a minimum of the functional

$$\Im_\alpha = \int_0^{t_f} [D^2 + 2æD\dot{\varphi} - \mu\dot{\varphi}^2] dt \tag{3.2.8}$$

and satisfy the conditions

$$\int_0^{t_f} M\dot{\varphi} dt = 0, \ |M| \le M_0, \ \dot{\varphi}(t) \ge 0, \tag{3.2.9}$$

when time t_f is not fixed.

Here

$$\begin{gathered} 2n = bJ^{-1}, \ N = N_1 J^{-1} \operatorname{sgn}\dot{\varphi}, \ D = ZD_1 J^{-1}, \\ M = M_1 J^{-1}, \ æ = \alpha k^2 Z^2 (2RJ)^{-1.} \end{gathered} \tag{3.2.10}$$

The integral limit (3.2.9) causes the problem to be isoperimetral [30,55].

It is suggested in references [53, 65], applied to the problem of vibration isolation, that it is possible to consider the force characteristic $M(\varphi)$ of a flexible element as the control term.

We choose φ as a new independent variable. Using the description $\dot{\varphi} = V$ we can represent the problem (3.2.7) to (3.2.9) in the form

$$\begin{gathered} dV^2 / d\varphi = 2(D + M - N - 2nV), \ V(0) = V(\varphi_f) = 0, \ V(\varphi) \ge 0, \\ dP_M / d\varphi = M, \ P_M(0) = P_M(\varphi_f) = 0, \\ \Im_a = \int_0^{\varphi_f} [D^2 V^{-1} + 2æD - \mu V] d\varphi, \ |\mathrm{M}| \le M_{0.} \end{gathered} \tag{3.2.11}$$

In what follows, the term V^2 will be taken as a new variable.

Using the maximum principle [90], the Hamiltonian for the system (3.2.11) becomes

$$\begin{gathered} H(V^2, \varphi, p_1, p_2, D, M) = \\ = \mu V - D^2 V^{-1} - 2æD + 2p_1(D + M - N - 2nV) + p_2 M. \end{gathered} \tag{3.2.12}$$

The equations for the conjugate variables are

$$dp_1 / d\varphi = -\partial H / \partial V^2 = (4p_1 n - D^2 V^{-2} - \mu) / 2V, \tag{3.2.13}$$

$$dp_2 / d\varphi = -\partial H / \partial P_M = 0, \tag{3.2.14}$$

and the conditions for optimisation are

$$D_* = \arg\max_D (2p_1 D - D^2 V^{-1} - 2æD), \tag{3.2.15}$$

$$M_* = \arg\max_M \{M(2p_1 + p_2) / |M| \le M_0\}.^1 \tag{3.2.16}$$

From the conditions (3.2.15) and (3.2.16) we find

$$D_* = (p_1 - æ)V, \tag{3.2.17}$$

$$M_* = M_0 \operatorname{sgn}(2p_1 + p_2) / 2p_1 + p_2 \ne 0. \tag{3.2.18}$$

From equation (3.2.14) it follows that $p_2 = \text{const}$, and using (3.2.17) we can write four solutions for equation (3.2.13)

[1] Here and below, the symbol '/' should be read as: 'satisfying the conditions' or 'when applying the conditions'.

$$
\begin{aligned}
p_{1a} &= æ + 2(n + \lambda \tanh \lambda\tau) / \\
/\mu &< 4n(n+æ),\ p_1 \in [æ + 2(n-\lambda), æ + 2(n+\lambda)],
\end{aligned} \tag{3.2.19}
$$

$$
\begin{aligned}
p_{1b} &= æ + 2(n + \lambda \coth \lambda\tau) / \\
/\mu &< 4n(n+æ),\ p_1 \notin [æ + 2(n-\lambda), æ + 2(n+\lambda)],
\end{aligned} \tag{3.2.20}
$$

$$
p_{1c} = æ + 2(n + \tau^{-1}) \ / \ \mu = 4n(n+æ), \tag{3.2.21}
$$

$$
p_{1d} = æ + 2(n - \nu \tan \nu\tau) \ / \ \mu > 4n(n+æ), \tag{3.2.22}
$$

where $\tau = \delta + t$, $\lambda = n[1 + æn^{-1} - \mu(4n^2)^{-1}]^{1/2}$, $\nu = j\lambda$, $j = (-1)^{1/2}$, t is current time, δ is a constant defined by the phase conditions (3.2.3) at the right hand end of the trajectory.

Fig. 3.2.1 shows the way in which the variables p_{1a}, p_{1b}, p_{1c} and p_{1d} vary.

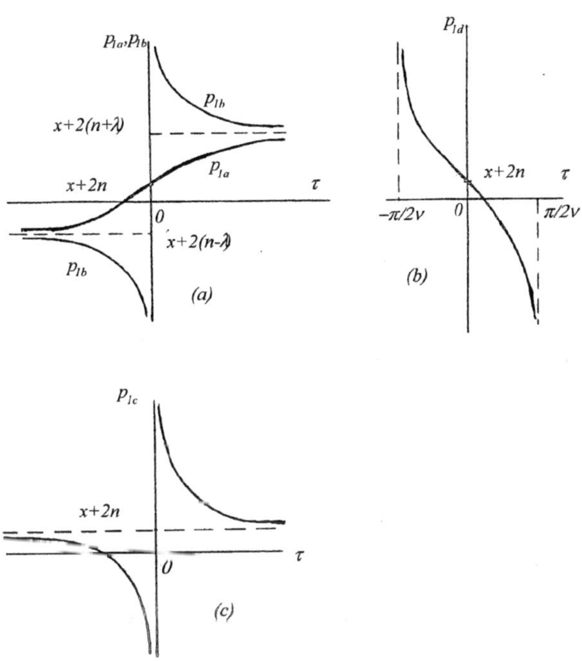

Fig. 3.2.1

The Hamiltonian (3.2.12) depends linearly on the reactive control M. Thus, if in the final part of the trajectory the argument of the function $\mathrm{sgn}(2p_1 + p_2)$ is identically zero then the required condition $H(\ldots M_*) \geq H(\ldots M)$ does not give information on the link between M_* and the phase variables and the Lagrangian multipliers. In this case, special control regimes are established [13, 34]. To verify the possibility that special control regimes occur, we will put $2p_1 + p_2 = 0$. Then $2p_1 = -p_2 = \mathrm{const}$ and $dp/d\varphi = 0$. From the condition that the Hamiltonian is constant, it appears that the special regimes with singular control will occur along portions of the trajectory, where the driven component is moving at constant velocity. From (3.2.13) and (3.2.17) and the requirement that $dp_1/d\varphi = 0$ it follows that the special regimes can occur only when $p_1 = æ + 2(n \pm \lambda)$. Fig. 3.2.1 shows that the control M_* can degenerate with the parameters p_{1a-1c} when $\tau \to \pm\infty$. The optimal control with parameter p_{1d} has no singularity.

For the non-linear problem (3.2.11) the maximum principle only gives necessary conditions for optimal control. From (3.2.19) to (3.2.22) it follows that the controls D_* and M_* can be non-unique. In addition, they are not capable of provide minimum values for the criterion functional in regions of the implied boundary conditions (3.2.3) and limited according to the control (3.2.9).

We shall verify whether the extreme controls (3.2.17) and (3.2.18) satisfy the minimum functional condition imposed by constraints (3.2.3), (3.2.7) and (3.2.9). From (3.2.17) it follows that when $V(0) = 0$ the active control $D_*(0) = 0$. Then to obtain motion $V(\varphi) > 0$ with finite time t_f it is necessary that $M_*(0) = M_0$ so that $M_0 > N$, to agree with (3.2.18). Thus, to satisfy the condition $P_M(\varphi_f) = 0$ the control M_* must have at least one switch point. It is also obvious that the constant δ must be finite. When $\delta \to \pm\infty$ in the initial phase of motion the control becomes singular with $M_* = N$. The drive is then in equilibrium and the boundary conditions can be satisfied only if $t_f \to \infty$.

It can be seen that the way in which the curve of p_{1a} varies does not ensure a switch of M_* from $+M_0$ to $-M_0$ for any values of τ and p_2. The curves for p_{1b} and p_{1c} when $\tau < 0$ correspond to the drive with $N < 0$, but when $V > 0$ it is in contradiction with the nature of dry friction forces in the drive. To verify the optimal control with p_{1b} for $\tau > 0$ and with p_{1d} for any value of τ, we will substitute (3.2.17) into the criterion functional. After appropriate transformation we find

$$\Im_a = \int_0^{\varphi_f} (p_1^2 - æ^2 - \mu) V d\varphi, \quad V \geq 0.$$

If, along the entire path the condition $p_1^2 > æ^2 + \mu$ is satisfied then $\Im_a \geq 0$. The functional reaches its minimum $\Im_a = 0$ when $V = 0$. Since the parameter p_{1b} when $\tau > 0$ ensures that the condition $p_1^2 > æ^2 + \mu$ is satisfied for all $\varphi \in [0, \varphi_f]$ and control (3.2.18) ensures that the driven component achieves maximum velocity along its entire path, then the value of the functional will not be a minimum, but a maximum.

If $p_1^2 = æ^2 + \mu$ for all $\varphi \in [0, \varphi_f]$ then the function is identically equal to zero. Such a motion regime is only possible with singular control.

If $p_1^2 < æ^2 + \mu$ for all $\varphi \in [0, \varphi_f]$ then $\Im_a < 0$ and the functional minimum is reached with the maximum velocity of motion (at every instant of time). Such a regime can be achieved by controls (3.2.17) and (3.2.18) with parameter p_{1d}. In this state the reactive control has one switch point, situated at the midpoint of interval $[0,\varphi_f]$.

This test analysis has shown that only one pair of optimal controls exists

$$D_* = 2(n - v\tan vt)V\,, \qquad (3.2.23)$$

$$M_* = M_0 \operatorname{sgn}(\varphi_f - 2\varphi)\,, \qquad (3.2.24)$$

which ensure the displacement of the driven component to the point φ_f with finite time t_f and minimum energy consumption by the driver. The controls found occur with weighting coefficients that satisfy the inequality $\mu > 4n(æ + n)$. For this, the criterion functional term that represents the drive operating speed dominates the energy terms.

Using (3.2.23) and (3.2.24) equation (3.2.7) for the optimal cyclic drive takes the form

$$\ddot{\varphi} + 2v\tan vt\,\dot{\varphi} + N = M_0 \operatorname{sgn}(\varphi_f - 2\varphi)\,. \qquad (3.2.25)$$

We can write the solutions of this equation for the initial conditions (3.2.3) [39, 47]

$$\varphi(t) = \frac{F_1}{2v^2\cos v\delta}\left[\sin vt \sin v(t+\delta) - vt\sin v\delta\right]\eta(t_1 - t) +$$

$$+\left\{\begin{array}{l} \varphi_f/2 + \dfrac{V_1}{2v\cos^2 v(\delta+t_1)}\left[\begin{array}{c}\sin v(t-t_1)\cos v(2\delta+t_1+t)+\\ +v(t-t_1)\end{array}\right] \\ -\dfrac{F_2}{2v^2\cos v(\delta+t)}\left[\begin{array}{c}\sin v(\delta+t)\sin v(t-t_1)-\\ -v(t-t_1)\sin v(\delta+t_1)\end{array}\right]\end{array}\right\}\eta(t-t_1) \qquad (3.2.26)$$

$$V(t) = \frac{F_1}{v\cos v\delta}\sin vt\cos v(\delta+t)\eta(t_1-t) +$$

$$+\left\{\begin{array}{l} V_1\cos^2 v(\delta+t)\cos^{-2} v(\delta+t_1) - \\ \qquad -\dfrac{F_2}{v\cos v(\delta+t_1)}\sin v(t-t_1)\cos v(\delta+t)\end{array}\right\}\eta(t-t_1).$$

where $V_1 = \dfrac{F_1}{v\cos v\delta}\sin vt_1\cos v(\delta+t_1)$, $F_1 = M_0 - N$, $F_2 = M_0 + N$, $\eta(x)$ is the symmetrical unit function [68]: $\eta(x) = 0$ when $x < 0$, $\eta(x) = 1/2$ when $x = 0$, and $\eta(x) = 1$ when $x > 0$, V_1 is the velocity of the driven component when the reactive control is switched in, t_1 is the instant of switching, found from the condition

$$\varphi_f = F_1[\sin v(\delta+t_1)\sin vt_1 - vt_1\sin v\delta](v^2\cos v\delta)^{-1}.$$

The ratio t_1/t_f increases as the value of dry friction N increases.

We now come to the solution of equation (3.2.7) for machine motion under the effect of the optimal reactive control (3.2.24) and active control (3.2.17) with parameters p_{1a}, p_{1b} and p_{1c}.

Using (3.2.17), (3.2.19) and (3.2.24), equation (3.2.7) can be written in the following form:

$$\ddot{\varphi}-2\lambda\tanh\lambda(\delta+t)\,\dot{\varphi}+N=M_0\,\mathrm{sgn}(\varphi_f-2\varphi)\,. \tag{3.2.27}$$

The solutions of (3.2.27) with the initial conditions (3.2.3) take the form

$$\varphi(t)=\frac{F_1}{2\lambda\cosh^2\lambda L}\begin{bmatrix}\sinh\lambda t\sinh\lambda(t+\delta)\\ -\lambda t\sinh\lambda\delta\end{bmatrix}\eta(t_1-t)+$$
$$+\left\{\begin{matrix}\varphi_f/2+\dfrac{V_1}{2\lambda\cosh^2\lambda(\delta+t_1)}\begin{bmatrix}\sinh\lambda(t-t_1)\cosh\lambda\left(2\delta+t_1+t\right)+\\ +\lambda(t-t_1)\end{bmatrix}-\\ -\dfrac{F_2}{2\lambda^2\cosh\lambda(\delta+t_1)}[\sinh\lambda(\delta+t)\sinh\lambda(t-t_1)-\\ -\lambda(t-t_1)\sinh\lambda(\delta+t_1)]\end{matrix}\right\}\times$$
$$\times\eta(t-t_1), \tag{3.2.28}$$

$$V(t)=\frac{F_1}{\lambda\cosh\lambda\delta}\sinh\lambda t\cosh\lambda(\delta+t)\eta(t_1-t)+$$
$$+\left\{\begin{matrix}V_1\cosh^2\lambda(\delta+t)\cosh^{-2}\lambda(\delta+t_1)-\\ -\dfrac{F_2}{\lambda\cosh\lambda(\delta+t_1)}\sinh\lambda(t-t_1)\cosh\lambda(\delta+t)\end{matrix}\right\}\eta(t-t_1),$$

where $V_1=\dfrac{F_1}{\lambda\cosh\lambda\delta}\sinh\lambda t_1\cosh\lambda(\delta+t_1)$, t_1 is the instant of switching, defined by the equation

$$\varphi_f=F_1[\sinh\lambda(\delta+t)\sinh\lambda t_1-\lambda t_1\sinh\lambda\delta](\lambda^2\cosh\lambda\delta)^{-1}\,.$$

For the parameter p_{1b} (3.2.20), using (3.2.3), (3.2.17) and (3.2.24), equations (3.2.7) can be written in the form

$$\ddot{\varphi}-2\lambda\coth\lambda(\delta+t)\dot{\varphi}+N=M_0\,\mathrm{sgn}(\varphi_f-2\varphi)\,. \tag{3.2.29}$$

with the solutions [49]

$$\varphi(t)=\frac{F_1}{2\lambda^2\sinh\lambda\delta}[\sinh\lambda t\cosh\lambda(t+\delta)-\lambda t\cosh\lambda\delta]\eta(t_1-t)+$$
$$+\left\{\begin{matrix}\varphi_f/2+\dfrac{V_1}{2\lambda\sinh^2\lambda(\delta+t_1)}\begin{bmatrix}\sinh\lambda(t-t_1)\cosh\lambda(2\delta+t_1+t)+\\ +\lambda(t-t_1)\end{bmatrix}-\\ -\dfrac{F_2}{2\lambda^2\sinh\lambda(\delta+t_1)}\begin{bmatrix}\cosh\lambda(\delta+t)\sinh\lambda(t-t_1)-\\ -\lambda(t-t_1)\cosh\lambda(\delta+t_1)\end{bmatrix}\end{matrix}\right\}\times \tag{3.2.30 a}$$
$$\times\eta(t-t_1),$$

$$V(t)=\frac{F_1}{\lambda\sinh\lambda\delta}\sinh\lambda t\sinh\lambda(\delta+t)\eta(t_1-t)+$$

$$+\left\{\begin{array}{l}V_1\sinh^2\lambda(\delta+t)\sinh^{-2}\lambda(\delta+t_1)-\\-\dfrac{F_2}{\lambda\sinh\lambda(\delta+t_1)}\sinh\lambda(t-t_1)\sinh\lambda(\delta+t)\end{array}\right\}\eta(t-t_1), \qquad (3.2.30\text{ b})$$

where $V_1=\dfrac{F_1}{\lambda\sinh\lambda\delta}\sinh\lambda t_1\sinh\lambda(\delta+t_1)$, t is the instant of switching, defined from the equation

$$\varphi_f=F_1[\cosh\lambda(\delta+t_1)\sinh\lambda t_1-\lambda t_1\cosh\lambda\delta](\lambda^2\sinh\lambda\delta)^{-1}.$$

For the parameter p_{1c} (3.2.21), using (3.2.3), (3.2.17) and (3.2.24), equation (3.2.21) takes the form

$$\ddot{\varphi}-\tau^{-1}\dot{\varphi}+N=M_0\,\mathrm{sgn}(\varphi_f-2\varphi) \qquad (3.2.31)$$

with the solution [49]

$$\varphi(t)=\frac{F_1}{4}\left[\tau^2\ln\frac{\tau^2}{\delta^2}-\tau^2+\delta^2\right]\eta(t_1-t)+$$

$$+\left\{\begin{array}{l}\varphi_f/2+\dfrac{V_1}{2|\delta+t_1|}\left[\tau^2-(\delta+t_1)^2\right]-\\-\dfrac{F_2}{4}\left[2\tau^2\ln\dfrac{|\tau|}{|\delta+t_1|}-\tau^2+(\delta+t_1)^2\right]\end{array}\right\}\eta(t-t_1), \qquad (3.2.32)$$

$$V(t)=F_1|\tau|\ln\frac{|\tau|}{|\delta|}\eta(t_1-t)+\left[\frac{V_1|\tau|}{|\delta+t_1|}-F_2|\tau|\ln\frac{|\tau|}{|\delta+t_1|}\right]\eta(t-t_1)$$

where $V_1=F_1|\delta+t_1|\ln\dfrac{|\delta+t_1|}{|\delta|}$, t_1 is found from the equation

$$\varphi_f=F_1\left[(\delta+t_1)^2\ln\frac{(\delta+t_1)^2}{\delta^2}-(\delta+t_1)^2+\delta^2\right]/2.$$

For the case $N=0$ the boundary condition (3.2.3) is satisfied in the system (3.2.7) by optimal controls (3.2.23) and (3.2.24) when the following relationships apply

$$t_f=2t_1=-2\delta,\ \varphi_f=V_1t_f/2,\ V_1=M_0\nu^{-1}\tan\frac{\nu t_f}{2}. \qquad (3.2.33)$$

Fig. 3.2.2 shows the phase portraits of $V(\varphi)$ and diagrams for the optimal controls $D(\varphi)$ and $M(\varphi)$ for various values of ν: i) $\nu=0.1$; ii) $\nu=1$; iii) $\nu=1.6$.

For the first half of the path the active control performs more work than that for the second half of the path. From Fig. 3.2.2, *b* it can be seen that for a sufficiently large value of ν, the work done during the first half of the path exceeds the total energy loss due to the friction during a complete cycle. Towards the end of the path the driver has to produce braking by acting either as a generator or switching to reverse. In the generator regime emf E in the driver is greater than the control voltage u (3.1.7), the current in the armature winding circuit and the active torque from the driver change sign. The mechanical energy

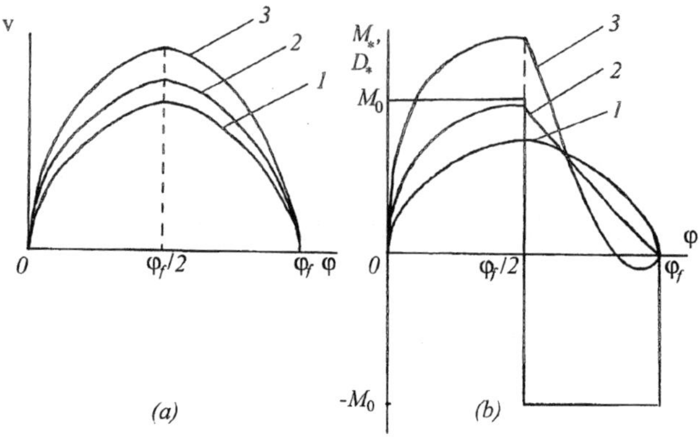

Fig. 3.2.2

of motion is converted to electromagnetic, and this returns into the electric circuit except for losses in the armature winding resistance [37].

The value of the functional (3.2.8) when $N=0$ and with controls (3.2.23) and (3.2.24) is defined by the equation

$$\Im_a = M_0^2[t_f \cos \nu t_f - \nu^{-1} \sin \nu t_f]/(\cos \nu t_f /2) < 0\,, \qquad (3.2.34)$$

where t_f is found from the expression $2\nu\varphi_f = M_0 t_f \tan(\nu t_f /2)$.

The implementation of precise optimal controls (3.2.23) and (3.2.24) in practice presents some technical difficulties. Optimal variation of the driver voltage change, which can be obtained from (3.2.23) when using (3.2.2) and (3.2.10), depends not only on the phase coordinate $\dot{\varphi}$, but also on the current time t

$$u = Zk\{1 + a æ^{-1}[n - \nu \tan \nu(\delta + t)]\}\dot{\varphi}\,. \qquad (3.2.35)$$

Control (3.2.35) can be implemented by a positive velocity feedback with a time-dependent gain factor.

The performance of a relieving device to form a relay type of optimal reactive control (3.2.24) is not a simple technical problem. The relieving device must include an energy store capable to produce a constant value torque M_0, switching clutches and also the mean of the control torque. As a store with constant output torque a clock-spring type of accumulator could be used [14, 47] (*Patent USSR 180045, IPC F16H33/02; Patent USSR 203411, IPC F16H33/06; Patent USSR 1132047, IPC F03G1/00*). Control clutches can operate as the switching devices.

Apart from the practical value in implementing controls (3.2.23) and (3.2.24) the principal importance in solving the problem of optimal dynamic synthesis of a cyclic drive consists of the fact that it becomes possible to evaluate available competitive variants of control and types of energy store.

Above all, from the solution (3.2.23) of the optimal problem it is possible to draw qualitative conclusion: in order that a cyclic manipulator drive should operate at high speed and economically it is necessary that it should be oscillatory and the control should be formed by the positive velocity feedback.

§3.3 Control of resonant drives with minimal energy consumption

The high dynamic quality of resonant drives occurs because inertia forces are generally in equilibrium with forces created by the elastic element energy store, and forces developed in the drive by the driver are used principally to compensate dissipative losses in the drive transmission. The problem of controlling the driver to provide the best method for compensating dissipative losses can be represented as a problem to generate optimal control of a resonant drive with mechanical energy store having fixed elastic characteristics when energy consumed by the control is minimised [96].

The dynamics of a resonant drive with a direct current independently supplied motor is described by the equations

$$\ddot{\varphi} + 2n\dot{\varphi} + M(\varphi) + N\,\mathrm{sgn}\,\dot{\varphi} = D,$$
$$D = \frac{Zk}{RJ}(u - Zk\dot{\varphi}), \qquad (3.3.1)$$

with boundary conditions

$$t = 0,\ \varphi = \varphi_s,\ \dot{\varphi} = \dot{\varphi}_s,$$
$$t = t_f,\ \varphi = \varphi_f,\ \dot{\varphi} = \dot{\varphi}_f, \qquad (3.3.2)$$

where $2n = bJ^{-1}$, $M(\varphi) = M_1(\varphi)J^{-1}$, $N = N_1 J^{-1}$, $D = ZQ_1 J^{-1}$ and the parameters have similar meanings as introduced in (3.2.1) to (3.2.3).

To generate a control to minimise the overall energy consumption of the driver represented by the functional (3.1.10), it is rational to state the problem with indeterminate cycle time t_f. In this case the operating speed of the drive is determined mainly by the condition of the energy store, but not by the control. Obviously, for the majority of resonant drive types the optimal cycle time will be finite. Exceptions will be those drives in which motion begins from a state of unstable equilibrium with zero initial velocity, and the elastic characteristic $M(\varphi)$ in the vicinity of this position is a smooth function. In particular, such a condition is provided by a stepped type of resonant module (Fig. 3.1.2). In this case the minimal energy consumption occurs when $t_f = \infty$.

We shall find the control to ensure the transition of the system (3.3.1) from the phase state ($\varphi_s, \dot{\varphi}_s$) to the new state ($\varphi_f, \dot{\varphi}_f$) with minimal energy losses over an indeterminate time t_f. We choose φ as the new independent variable. For clarity we will assume that at the start of the motion cycle $\dot{\varphi} \geq 0$. Using the variable $\dot{\varphi} = V$, (3.3.1) can be put in the form

$$dV^2 / d\varphi = -2M(\varphi) - 4nV - 2N + 2D,$$
$$V(\varphi_s) = V_s,\quad V(\varphi_f) = V_f,\quad V \geq 0. \qquad (3.3.3)$$

Minimised functional (3.1.10) takes the form

$$\Im_b = \int_{\varphi_s}^{\varphi_f} (D^2V^{-1} + 2æD)d\varphi \,. \tag{3.3.4}$$

Here $æ$ is analogous to (3.2.10) and takes the form

$$æ = aZ^2k^2(2RJ)^{-1},$$

where a is a non-negative coefficient. When $a = 0$ the problem leads to the minimisation of heat losses in the electrical parts of the drive; when $a = 1$ the entire energy losses of the driver are minimised; when $a \to \infty$ the mechanical work performed by the driver is minimised.

To solve equations (3.3.3) and (3.3.4) we shall use the maximum principle [90]. The Hamiltonian of the system takes the form

$$H(V^2, \varphi, p, D) = -2p[M(\varphi) + 2nV + N - D] - D^2V^{-1} - 2æD \tag{3.3.5}$$

where the Lagrangian multiplier satisfies the equation $dp/d\varphi = \partial H/\partial V^2$, or

$$dp/d\varphi = (4np - D^2V^{-2})(2V)^{-1} \,. \tag{3.3.6}$$

The optimal control D_* satisfies the maximum condition

$$D_* = \arg\max_D H(\ldots D) = \arg\max_D \{2pD - D^2V^{-1} - 2æD\} \,.$$

From this it follows that

$$D_* = (p - æ)V \,. \tag{3.3.7}$$

Equation (3.3.7) is similar in structure to expression (3.2.17) obtained earlier.

The substitution of (3.3.7) into (3.3.6) leads to the separation of possible controls D_* into two types depending on the factor p. When $p \in [æ + 2(n - \lambda),\ æ + 2(n + \lambda)]$ then control D_M (known as "weak" control) occurs, is given by

$$D_M = (n + \lambda \tanh \lambda\tau)2V \,, \tag{3.3.8}$$

where $\lambda = n(1 + æn^{-1})^{1/2}$, $\tau = \delta + t$, δ is a constant found from the phase condition at the right hand end of the trajectory.

When $p \notin [æ + 2(n - \lambda),\ æ + 2(n + \lambda)]$ we have control D_B (known as "strong" control) given by

$$D_B = (n + \lambda \coth \lambda\tau)2V \,. \tag{3.3.9}$$

Fig. 3.3.1 shows the way that the term ($p - æ$) varies with time τ.

Hence, optimal control will be that which is proportional to the velocity of motion of the driven component. The term ($p - æ$) depends on current time. From Fig. 3.3.1 and equation (3.3.8) it can be seen that for small values of λ, which is representative of actual drives with small dissipation, the term ($p - æ$) depends insignificantly on time.

Using (3.3.8) and (3.3.9) it is possible to write the initial equation of motion for weak and strong control for a resonant drive as

$$\begin{gathered} \varphi - 2\lambda W(\tau)\dot{\varphi} + M(\varphi) + N = 0, \\ \varphi(0) = \varphi_s,\ \dot{\varphi}(0) = \dot{\varphi}_s,\ \varphi(t_f) = \varphi_f,\ \dot{\varphi}(t_f) = \dot{\varphi}_f, \end{gathered} \tag{3.3.10}$$

where $\tau = \delta + t$, $W(t) = \begin{cases} \tanh \lambda\tau / p \in [æ + 2(n-\lambda),\ æ + 2(n+\lambda)], \\ \coth \lambda\tau / p \notin [æ + 2(n-\lambda),\ æ + 2(n+\lambda)]. \end{cases}$

In the general case for a nonlinear elastic characteristic $M(\varphi)$, equation (3.3.10) is solved numerically.

For the cyclic resonant drive [4] shown in Fig. 3.1.1 with the linear elastic characteristic

$$M(\varphi) = \omega^2 \varphi , \tag{3.3.11}$$

where $\omega^2 = CJ^{-1}$ and C is the stiffness of the elastic elements, equation (3.3.10) has an analytic solution [57].

The equation for weakly controlled motion when $\dot{\varphi} \geq 0$ takes the form

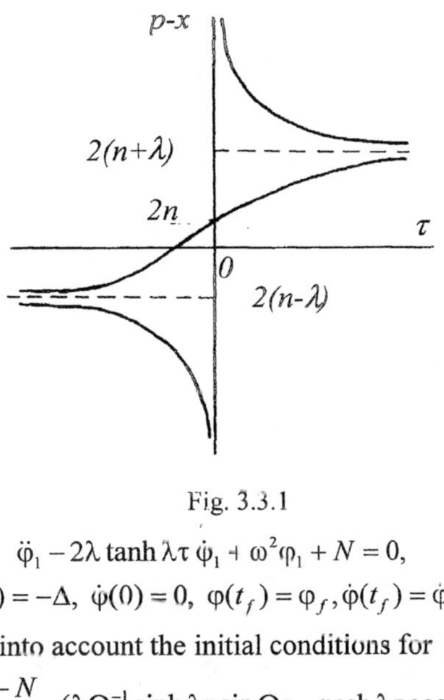

Fig. 3.3.1

$$\begin{gathered} \ddot{\varphi}_1 - 2\lambda \tanh \lambda\tau\, \dot{\varphi}_1 + \omega^2 \varphi_1 + N = 0, \\ \varphi(0) = -\Delta,\ \dot{\varphi}(0) = 0,\ \varphi(t_f) = \varphi_f, \dot{\varphi}(t_f) = \dot{\varphi}_f . \end{gathered} \tag{3.3.12}$$

The solution, taking into account the initial conditions for $\lambda\omega^{-1} < 1$, is as follows:

$$\begin{gathered} \varphi_1(t) = \frac{\Delta\omega^2 - N}{\omega^2 \cosh \lambda\delta_1} (\lambda\Omega^{-1} \sinh \lambda\tau \sin \Omega\tau - \cosh \lambda\tau \cos \Omega\tau) - N\omega^{-2}, \\ \dot{\varphi}_1(t) = \frac{\Delta\omega^2 - N}{\Omega \cosh \lambda\delta_1} \cosh \lambda\tau \sin \Omega\tau, \end{gathered} \tag{3.3.13}$$

where $\Omega^2 = \omega^2 - \lambda^2$.

Using (3.3.13), equation (3.3.8) for the change of weak control can be written in the form

$$D_M = \frac{2\sqrt{æn}(\Delta\omega^2 - N)}{\Omega \cosh \lambda\delta_1} \sinh(\lambda\tau + \gamma) \sin \Omega\tau , \tag{3.3.14}$$

where $\gamma = \tanh^{-1}(1 + æn^{-1})^{1/2}$.

In the case of strong control, which can occur in motion systems with $\dot{\varphi} > 0$ or $\dot{\varphi} < 0$, we find

$$\ddot{\varphi}_2 - 2\lambda \coth \lambda\tau \, \dot{\varphi}_2 + \omega^2 \varphi_2 + N \operatorname{sgn} \dot{\varphi}_2 = 0 , \tag{3.3.15}$$

where when $\lambda\omega^{-1} < 1$ we can write

$$\begin{gathered} \varphi_2(t) = \frac{\Delta\omega^2 - N \operatorname{sgn} \dot{\varphi}_2}{\omega^2 \cosh \lambda\delta_2} \begin{pmatrix} \lambda\Omega^{-1} \cosh \lambda\tau \sin \Omega t - \\ - \sinh \lambda\tau \cos \Omega t \end{pmatrix} - N\omega^{-2} \operatorname{sgn} \dot{\varphi}_2, \\ \dot{\varphi}_2(t) = \frac{\Delta\omega^2 - N \operatorname{sgn} \dot{\varphi}_2}{\Omega \cosh \lambda\delta_2} \sinh \lambda\tau \sin \Omega t, \end{gathered} \tag{3.3.16}$$

The variation of driver torque can be found after substituting (3.3.16) into (3.3.9)

$$D_B = \frac{2(æn)^{-1}(\Delta\omega^2 - N \operatorname{sgn} \dot{\varphi}_2)}{\Omega \sinh \lambda\delta_2} \cosh(\lambda\tau + \gamma) \sin \Omega t , \tag{3.3.17}$$

Equations (3.3.8) and (3.3.9) represent optimal controls written as a model of synthesis, and equations (3.3.14) and (3.3.17) are written as the model for a program to produce the same controls in time.

From the phase condition for $\dot{\varphi}(\varphi_f)$ at the right hand end of the trajectory t_f and δ can be found by simultaneously solving equations (3.3.13) or (3.3.16). For cyclic drives the main concern is to satisfy the condition $\dot{\varphi}_f = 0$. It is obvious that in both cases $\dot{\varphi}_f = 0$ at the instant when $t_f = \pi / \Omega$. As a result, depending on the value of δ and the type of control, the driven component will take up different positions φ_f. From expressions (3.3.13) and (3.3.16) we can write accordingly

$$\varphi_{f1} = (\Delta - N\omega^{-2}) \cosh(T + \lambda\delta_1)(\cosh \lambda\delta_1)^{-1} - N\omega^{-2} , \tag{3.3.18}$$

$$\varphi_{f2} = (\Delta - N\omega^{-2} \operatorname{sgn} \dot{\varphi}) \sinh(T + \lambda\delta_2)(\sinh \lambda\delta_2)^{-1} - N\omega^{-2} \operatorname{sgn} \dot{\varphi} . \tag{3.3.19}$$

where $T = \pi\lambda\Omega^{-1}$.

We can express the constant δ in terms of φ_f

$$\delta_1 = \lambda^{-1} \tanh^{-1} Y, \quad \delta_2 = \lambda^{-1} \coth^{-1} Y , \tag{3.3.20}$$

where $Y = (\varphi_f \omega^2 + N \operatorname{sgn} \dot{\varphi}) \sinh T (\Delta\omega^2 - N \operatorname{sgn} \dot{\varphi})^{-1} - \coth T$.

Fig. 3.3.2 shows the graph of the coordinate $\varphi_f = \varphi(\pi\Omega^{-1})$, the dwell point of the driven component, in terms of the constant δ. The curve for φ_f crosses the ordinate axis at the point $\varphi_0 = (\Delta - N\omega^{-2}) \cosh T - N\omega^{-2}$ and is asymptotic to the lines $\delta = 0$,

$$\varphi^+ = (\Delta - N\omega^{-2})(\cosh T + \sinh T) - N\omega^{-2} , \tag{3.3.21}$$

$$\varphi^- = (\Delta - N\omega^{-2})(\cosh T - \sinh T) - N\omega^{-2} . \tag{3.3.22}$$

The choice of δ is influenced by the required value of φ_f. To move the

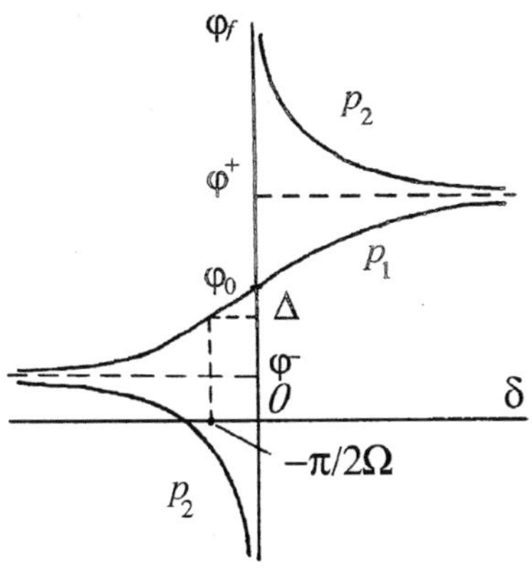

Fig. 3.3.2

driven component to the position $\varphi_f = +\Delta$, symmetrical to the starting point when $N = 0$, it is necessary to provide weak control with

$$\delta = -\pi(2\Omega)^{-1}. \tag{3.3.23}$$

Analysis of the behaviour of the solution shows that, with dry friction N, weak control ensures that the driven component arrives at the point $\varphi_f = +\Delta$ when the value of δ_1 is large compared to equation (3.3.23). As dry friction increases to the value

$$N^+ = \Delta\omega^2(\cosh T + \sinh T - 1)(\cosh T + \sinh T + 1)^{-1}, \tag{3.3.24}$$

weak control can not ensure that the condition $\varphi_f = +\Delta$ is satisfied even with an infinitely large constant δ_1. Weak control must be replaced by strong control in which a large value of N corresponds to a small value of δ_2.

If the value of N in equation (3.3.10) is considered as a combined torque due to the effect of a dry friction force and the weight of the load carried then the situation can arise when $N < 0$. For example, this can happen when a load that is greater than the dry friction force reduces. Then, to satisfy the condition $\varphi_f = +\Delta$ the driver must work in a braking regime. Under these conditions the optimal will be a constant δ less than that of (3.3.23). For a value of N less than the value

$$N^- = \Delta\omega^2(\cosh T - \sinh T - 1)(\cosh T + \sinh T + 1)^{-1}, \tag{3.3.25}$$

strong control (3.3.17) is required. For increasingly negative values of N appropriately less negative values of δ_2 are required.

With weak control it can occur that over the initial part of the trajectory the emf in the driver will be greater than the control voltage. As a result the current in the armature windings and the electromagnetic torque of the driver both change sign. The driver will operate as a generator in parallel with the circuit (recuperative braking). In this regime the mechanical energy due to the active operating torque $M(\varphi)$ is transformed into electromagnetic energy which is transmitted to the electricity supply network in an amount adjusted for the energy lost in the resistance of the armature circuit [102]. The appearance of the generator regime $D_M < 0$ is due to values of $æ, n$ and ω which satisfy the inequality $\lambda n^{-1} \tanh \lambda \delta_1 < -1$, which is characteristic of the relationship of the coefficients with $\dot{\varphi}$ in equations (3.3.1) and (3.3.10) when $t = 0$. From the inequality it can be seen that for small values of $æn^{-1}$ recuperative braking occurs in the mechanical system as the damping $n\omega^{-1}$ increases. When the condition $a = 0$ occurs, which corresponds to the initial problem of minimising only the heat losses in the driver, a generator regime with optimal control does not occur.

Fig. 3.3.3 shows the variation of torque $D_M(\varphi)$ and the optimal phase trajectories $V(\varphi)$ with weak control of the drive with various transmission and electric motor parameters. Curves *1* correspond to the parameters $n\omega^{-1} = 0.1;\ a = 0$, curves *2* to $n\omega^{-1} = 0.1;\ æn^{-1} = 25$; curve 3 to $n\omega^{-1} = 0.8;\ a = 0$; curve 4 to $n\omega^{-1} = 0.8;\ æn^{-1} = 0.4$. In the figure it can be seen that the optimal control (3.3.14) causes a symmetrical phase trajectory. As the damping increases the maximum optimal torque is displaced from the equilibrium position to the right hand boundary of the trajectory.

In practice, to produce the optimal control (3.3.7) is technologically difficult since the coefficient associated with velocity depends on the actual time. In

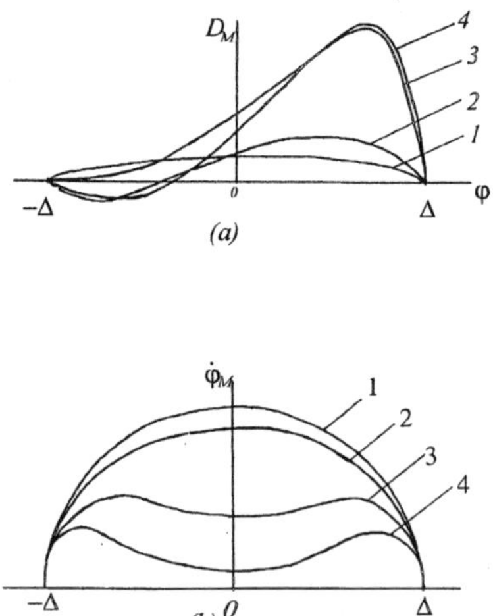

Fig. 3.3.3

addition, it is not always possible to arrange to recover energy, for example, in cases using thyristor devices [92]. Direct current motors with independent excitation are controlled by varying the armature voltage. Taking account of

(3.3.1) and (3.2.10) we can write (3.3.7) for the relative control voltage

$$u = Zk\{1 + a\text{æ}^{-1}[n + \lambda W(\tau)]\}\dot{\varphi}\,, \tag{3.3.26}$$

where W is found from (3.3.10). Note that the structure of control (3.3.26) is similar to (3.2.35) to secure the required drive operating speed with minimal energy consumption.

To avoid the driver operating as a generator it is possible to choose the transmission ratio Z, in which the parameter æ is defined in (3.2.10), to ensure that following condition is satisfied

$$(1 + \text{æ}/n)^{1/2}\tanh\left[\lambda\delta_1(1 + \text{æ}/n)^{1/2}\right] > -1\,.$$

In the general case, optimal control of cyclic resonant drive uses positive feedback link from a tachogenerator with an appropriate amplification coefficient

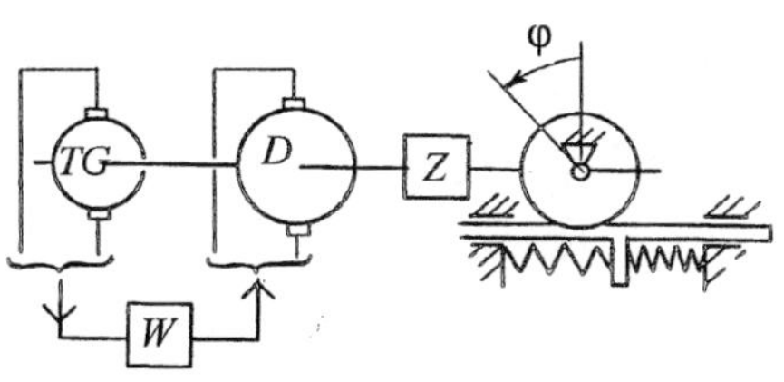

Fig. 3.3.4

$W(\tau)$ (Fig. 3.3.4).

The value of the energy functional (3.3.4) due to control (3.3.14) and assuming that $N = 0$, $\varphi_f = +\Delta$ and $\dot{\varphi}_f = 0$ is found from the following expression:

$$\Im_b = 4\lambda\Delta^2\omega^2\tanh\frac{\pi\lambda}{2\Omega}\,. \tag{3.3.27}$$

§3.4 Operating speed of cyclic resonant drives

One of the basic advantages of resonant systems is their high operating speed. As an example of a cyclic resonant module (Fig. 3.1.1) we will construct the optimal operating speed control for a drive with electric motor that has unsymmetrical motion of the module arm. The lack of symmetry is due to variations in the mass transported and in the coordinate of the positioning point in the forward and reverse directions. The driver is not inertia-free and the arm must approach the fixed edge position with moderate velocity [20].

The dynamics of the resonant module moving between the fixed positioning points is described by the following differential equations

$$\begin{gathered} J\ddot{\varphi} + b\dot{\varphi} + C\varphi = ZQ, \\ \tau\dot{Q} + Q = kR^{-1}u - Zk^2R^{-1}\dot{\varphi}, \end{gathered} \tag{3.4.1}$$

where $J = J_0 + J_1(1 + \operatorname{sgn}\dot{\varphi})/2$; J_0 and J_1 are moments of inertia of the arm and the transported mass, respectively; τ is the electromagnetic time constant of the driver (1.3.6); remaining terms are as taken in (1.3.6) and (3.2.1) to (3.2.3).

We shall develop the equation for $u(t)$, $|u| \le U_0$, to produce the optimal operating speed in the periodic system (3.4.1) to satisfy the given conditions of approach to the latch.

We shall deal separately with forward ($\dot{\varphi} > 0$) and reverse motion ($\dot{\varphi} < 0$) to create controls to optimise the operating speed, for each of these sections.

Taking into account that the dissipative forces and hysterisis in the system are small and introducing a small parameter $\varepsilon << 1$, we can rewrite the equation for forward motion ($\dot{\varphi} > 0$) in the form

$$\ddot{\varphi} + 2\varepsilon n\dot{\varphi} + \omega^2\varphi = \varepsilon Z_1 D,$$
$$\varepsilon\dot{D} = \tau^{-1}(-D + \alpha u - \gamma Z_1\dot{\varphi}), \tag{3.4.2}$$

where $\omega^2 = C/(J_0 + J_1)$, $2\varepsilon n = b/(J_0 + J_1)$, $Z_1 = Z/(J_0 + J_1)$, $\alpha = k/R$ $\gamma = (J_0 + J_1)k^2/R$, $\varepsilon D = Q$.

Forward motion begins from rest when the arm coordinate is determined by the position of the lower latch

$$\varphi(0) = \varphi_s, \quad \dot{\varphi}(0) = 0, \tag{3.4.3}$$

and finishes its travel at the upper latch with the required velocity

$$\varphi(t_f) = \varphi_f, \quad \dot{\varphi}(t_f) = V_f. \tag{3.4.4}$$

We shall find the control $u(t)$, $|u| \le U_0$, optimised for operating speed (3.1.6) for the system (3.4.2) with boundary conditions (3.4.3) and (3.4.4).

In place of the second equation (3.4.2) we can reduce the order of the system, putting as a first approximation

$$D = \alpha u - \gamma Z_1\dot{\varphi}, \tag{3.4.5}$$

and find the control $u^0(t)$, optimised for operating speed for the system with one degree of freedom

$$\ddot{\varphi} + 2\varepsilon\delta\dot{\varphi} + \omega^2\varphi = \varepsilon Z_2 u, \tag{3.4.6}$$

with boundary conditions (3.4.3) and (3.4.4), where $\delta = n + Z_1^2\gamma/2$, $Z_2 = Z_1\alpha$.

The basis for this approach is given in the literature [87]. It can be shown that when the system (3.4.2) is substituted by equations (3.4.5) and (3.4.6), the error in finding the optimal trajectory $[\varphi(t), D(t)]$ will be of $\Theta(\varepsilon)$ order, and the error in the finding operating speed will be of order $\Theta(\varepsilon^2 t)$. The notation $\Theta(\varepsilon)$ stands for terms of order ε.

The presence of the small parameter in the right hand side of equation (3.4.6) implies that the system operates in a resonant regime and only a low-power drive is required to compensate dissipative losses. Optimisation requirements determine the character of the arrangement as a resonant drive.

To clarify the features of resonant motion we transform equation (3.4.6) into a standard form. Introducing new variables of "amplitude-phase" according to the expressions

$$\varphi = A\cos\psi, \quad \dot{\varphi} = -A\sin\psi, \tag{3.4.7}$$

we obtain

$$\begin{aligned} \dot{A} &= -\varepsilon\omega^{-1}(Z_2 u + 2\delta\omega A \sin\psi)\sin\psi, \\ \dot{\psi} &= \omega - \varepsilon(\omega A)^{-1}(Z_2 u + 2\delta\omega A \sin\psi)\cos\psi. \end{aligned} \tag{3.4.8}$$

Considering ψ as a new independent variable and neglecting the terms of $\Theta(\varepsilon^2)$ order, we obtain (3.4.8) in the form

$$\begin{aligned} dA/d\psi &= -\varepsilon\omega^{-2}(Z_2 u + 2\delta\omega A \sin\psi)\sin\psi, \\ dv/d\psi &= \varepsilon\omega^{-2}A^{-1}(Z_2 u + 2\delta\omega A \sin\psi)\cos\psi, \end{aligned} \tag{3.4.9}$$

where $v = \omega t - \psi$. The boundary conditions for A and v and the phase ψ_f at the end of the process are established from the boundary conditions (3.4.3) and (3.4.4):

$$\begin{aligned} \psi(0) = 0,\ \psi_f &= \pi \tan^{-1}(V_f/\omega\varphi_f),\ v(0) = 0, \\ v(\psi_f) = \omega t_f - \psi_f,\ A(0) &= \psi_s,\ A(\psi_f) = \varphi_f^2 + (V_f/\omega)^2. \end{aligned} \tag{3.4.10}$$

From the expression for $v(\psi_f)$ it can be seen that the problem of minimising time of motion t_f after substituting the variables (3.4.7) leads to minimisation of the phase of the ending of the process

$$\arg\min_{|u|\le U_0} t_f \equiv \arg\min_{|u|\le U_0} v(\psi_f) \tag{3.4.11}$$

To find the solution we shall use the method of successive approximations [87]:

$$A = \varphi_s + \varepsilon A_1 + \varepsilon^2 \ldots,\quad v = \varepsilon v_1 + \varepsilon^2 \ldots, \tag{3.4.12}$$

where the functions A_1, v_1, and u_0 satisfy the equations

$$\begin{aligned} dA_1/d\psi &= -\omega^{-2}(Z_2 u + 2\delta\omega\varphi_s \sin\psi)\sin\psi, \\ dv_1/d\psi &= \omega^{-2}\varphi_s^{-1}(Z_2 u + 2\delta\omega\varphi_s \sin\psi)\cos\psi. \end{aligned} \tag{3.4.13}$$

The effect of (3.4.7), (3.4.10) and (3.4.12) for agreement between the boundary and initial conditions can be written in the form

$$\begin{aligned} \varphi_f = \varphi_s \cos\psi_f + \varepsilon æ,\quad A(\psi_f) = \varphi_s - \varepsilon\rho, \\ v(\psi_f) = \varepsilon\lambda,\quad t_f = \omega^{-1}(\psi_f + \varepsilon\lambda). \end{aligned}$$

Further on we shall limit how the arm approaches the latch to the minimum velocity $V_f = \varepsilon^{1/2} V_*$. Then

$$\psi_f = \pi - \varepsilon^{1/2}(V_*/\omega\varphi_f),\quad \varphi_f = -\varphi_s + \varepsilon æ. \tag{3.4.14}$$

In agreement with the assumed accuracy

$$\begin{aligned} A_1(0) - 0,\ A_1(\psi_f) &= -æ + (V_*/\omega)^2/(2\varphi_s) = -\rho, \\ v_1(0) = 0,\ v_1(\psi_f) &= \lambda. \end{aligned} \tag{3.4.15}$$

Using equation (3.4.13) we will write conditions (3.4.15) in the form of isoperimetric expressions [30,55]

$$A_1 = -\omega^{-2}\int_0^{\psi_f}[Z_2 u + 2\delta\omega\varphi_s \sin\psi]\sin\psi\, d\psi, \tag{3.4.16}$$

$$v_1 = \omega^{-2}\varphi_s^{-1}\int_0^{\psi_f}[Z_2u + 2\delta\omega\varphi_s\sin\psi]\cos\psi d\psi\,. \tag{3.4.17}$$

The condition for minimising objectives (3.1.6), (3.4.13) to (3.4.15) takes the form

$$u_* = \arg\min_{|u|\le U_0}\{v_1 + qA_1\}\,, \tag{3.4.18}$$

where q is a constant multiplier. Using (3.4.16) and (3.4.17) we find

$$u_* = \arg\min_{|u|\le U_0}\int_0^{\psi_f} Bu\sin(\psi - \psi_0)d\psi\,, \tag{3.4.19}$$

where $B = (\varphi_s^{-2} + q^2)^{1/2}$, $\tan\psi_0 = (q\varphi_s)^{-1}$.

From (3.4.19) it follows that

$$u_* = -U_0\,\mathrm{sgn}\sin(\psi - \psi_0)\,, \tag{3.4.20}$$

where the shift of phase ψ_0 fixes the instant of control transition.

The transition point ψ_0 is found from (3.4.16). The phase at the end of motion (end of control) can be found from (3.4.17).

Substituting the control from (3.4.20) into (3.4.16) and (3.4.17) and taking account of (3.4.14) and (3.4.15), we obtain the equation to find the transition point and the phase at the end of motion in the forward direction:

$$\begin{gathered}\rho\omega^2 = -2Z_2U_0\cos\psi_0 + \pi\delta\omega\varphi_s,\\ \lambda\omega^2\varphi_s = 2Z_2U_0\sin\psi_0.\end{gathered} \tag{3.4.21}$$

From (3.4.21) it follows immediately that

$$\begin{gathered}\cos\psi_0 = -r/(2Z_2U_0)^{-1},\\ \lambda = (4Z_2^2U_0^2 - r^2)^{1/2}/(\varphi_s\omega^2),\end{gathered} \tag{3.4.22}$$

where $r = \omega^2(\rho - \pi\delta\varphi_s/\omega)$.

We shall evaluate the extreme capacity of the control. Setting $\varphi_s = -\Delta$, then from (3.4.22) it can be seen that in all cases $\lambda < 0$. This expression shows that whatever the parameters in a resonant drive, then control (3.4.20) accelerates motion of the arm by comparison to the original uncontrolled symmetrical conservative system $(\delta = u = V_f = 0,\ \varphi_f = -\varphi_s = +\Delta)$.

Control of the original symmetrical conservative system $(æ = V_f = \delta = \rho = 0,$ $\varphi_f = +\Delta)$ occurs with phase transition $\psi_0 = \pi/2$ (Fig. 3.4.1).

In cases when one of the system parameters or several of them $(æ,\ V_f,\ \delta,\ \rho)$ simultaneously have nonnegative values the phase of control transition is shifted to the right hand boundary $\psi_0^+ > \pi/2$ (Fig. 3.4.1).

With unsymmetrical limits when $æ < 0$ the phase control transition is reduced to $\psi_o^- < \pi/2$ (Fig. 3.4.1).

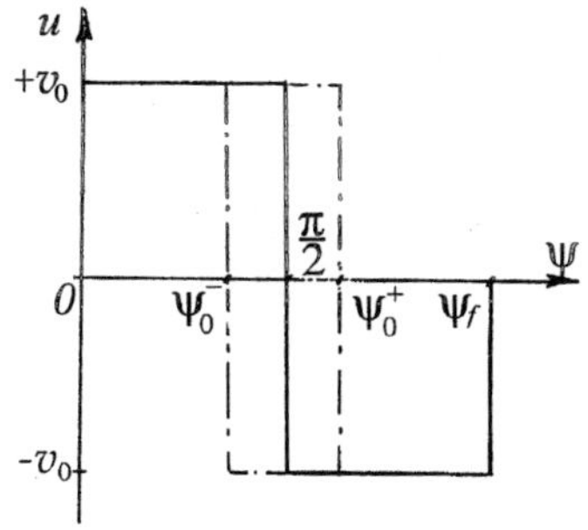

Fig. 3.4.1

Fig. 3.4.2 shows the expression (3.4.22) connecting the transition phase ψ_0 and the parameter r. The transition point comes closer to the end limit of motion $\psi_0 \to \pi$ whenever friction in the system, δ, and the regulated velocity of approach to the latch, V_f, are higher and the more unsymmetrical the value of $æ$, the latch disposition.

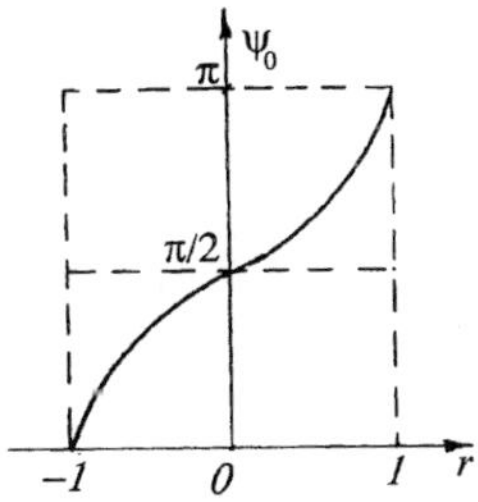

Fig.3.4.2

Fig 3.4.3 for the original symmetrical conservative system ($æ = \delta = 0,\ V_f > 0$) shows the smoothing effect the electromagnetic time constant τ of the driver has on the step change in the control voltage u.

Similar calculations can be carried out to study reverse motion ($\dot{\varphi} < 0$) of the

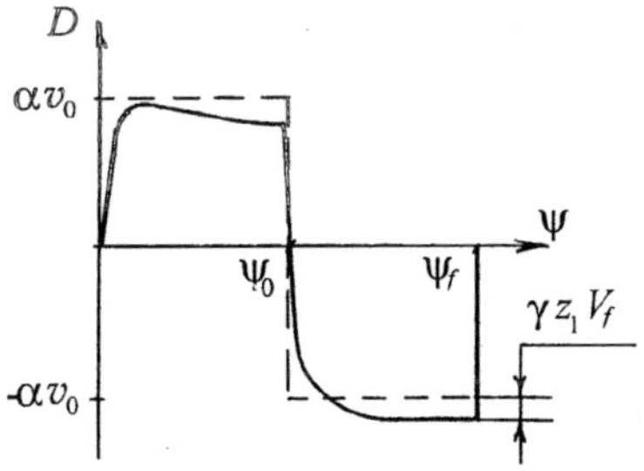

Fig. 3.4.3

arm. For this it is assumed that $\varphi(0)=\varphi_f$, $\dot{\varphi}(0)=0$, $\varphi(t_f)=\varphi_s$, $\dot{\varphi}(t_f)=V_f$, $J_1=0$. Taking account of the new values of the system parameters all the results for the control and character of forward motion are also admissible for reverse motion of the arm of a cyclic resonant module.

Hence, comparing the optimal operating speed results with the control scheme studied in §2.3, we find that using additional real characteristics of systems leads to the refinement of motion controls and assesses the possible limits for resonant manipulator systems.

Calculation shows that using disabling intervals to control the driver impairs the operating speed by comparison to the optimal value by the value of $\Theta(\varepsilon)$ order.

§3.5 Optimal control with minimal heat generation

Depending on the operating regime resonant drives can be divided into cyclic (Fig. 3.1.1) or stepped (Fig. 3.1.2). The operation of a cyclic resonant drive is implemented by oscillating the driven component relative to the stable position with latches at the extreme positions. The operating speed for this is determined by the natural frequency of the transmission equipped with mechanical energy store. With low dissipative losses in the drive and displacement occurring in the horizontal plane, and using a resonant system [12], a low-power driver can be used. The system remains weakly controlled [34, 87]. It has almost free motion, with weak control capable to change the total mechanical energy of the system by a small amount over a finite time interval [19].

The operation of a stepped resonant drive occurs when driven crank rotates from an unstable equilibrium position where it has been locked to the same position after manipulation through the required number of steps (*Patent USSR 1357219, IPC B25J9/00*). One turn of the crank corresponds to a displacement of one step for the driven component. Control of the motor during one step differs from the control when operating several steps. In multi-stepped motion regimes the driver maintains constant total mechanical energy, compared to operation when dissipative losses are compensated. To achieve high operating speeds with low energy consumption it is necessary that the value of the control in the first and final steps are significant and control must be weak over the intermediate steps. The search for optimal control during the first and final steps is carried out by the method explained in §3.6 [21].

Thus, weak control ensures motion of a cyclic resonant drive and is used for the intermediate steps of motion for stepped resonant drives.

Under intensive reciprocating motion of a cyclic resonant drive with short-term dwell at the boundary locations and also in regimes with multiple step motion, the motor is under load for much of the time. As a result a vital problem is to reduce the heat energy in the motor windings. In addition, it is also important to maintain a stable operating regime for the drive. These considerations form the basis of the two optimisation problems, to find weak control to ensure stable motion of the driven component with low heat emission in the armature windings of the motor.

The first problem takes the form of Lagrange's problem in which the function (3.1.11) is minimised with restricted control and fixing of t_f and V_f - the time and velocity of approach to the latches [19].

The second problem is generated by combining (3.1.11) and (3.1.14) for a functional (Bolza's problem) with indeterminate time t_f and with no limitation on the controlling voltage [22].

§3.5.1 Lagrange's problem. During motion between the extreme positions of arm *1* (Fig. 3.1.1) the dynamics of a cyclic resonant module with electric drive (1.3.6) can be described by the following differential equations

$$J\ddot{\varphi} + b\dot{\varphi} + C\varphi = ZQ,$$
$$\tau\dot{Q} + Q = kR^{-1}u - Z\chi(u)k^2R^{-1}\dot{\varphi}, \qquad (3.5.1)$$

and the boundary conditions

$$\varphi(0) = \varphi_s,\ \ \dot{\varphi} = 0,\ \ Q(0) = 0,$$
$$\varphi(t_f) = \varphi_f,\ \ \dot{\varphi}(t_f) = V_f. \qquad (3.5.2)$$

Here φ is the angle of rotation of the arm, measured from the position of equilibrium of the spring with stiffness C, Z is the transmission ratio of the gear-set, t_f and V_f are the fixed instant of time and controlled velocity of approach of the arm to the latch. In the motor characteristic, τ is the electromagnetic time constant, coefficient $\chi(u)$ is introduced to describe the switching-off of the controlling voltage when the current in the armature and, as a result, the torque on the driver instantaneously decay

$$\chi(u) = \begin{cases} 1, & u \neq 0, \\ 0, & u = 0. \end{cases}$$

The remaining meanings are similar to those taken in (1.3.6) and (3.2.1) to (3.2.3).

We shall construct an expression for the control voltage $u(t)$, $0 \le u \le U_0$ occurring in the system (3.5.1) for a resonant regime under the condition of a minimum functional to characterise heat generation (3.1.11) in the motor

$$\Im_{d1} = \int_0^{t_f} Q^2(t)dt\,, \qquad (3.5.3)$$

Taking into account that the driver time constant τ is small and bearing in mind the resonant character of the motion sought, we can put (3.5.1) into the form

$$\ddot{\varphi} + 2\varepsilon n\dot{\varphi} + \omega^2\varphi = \varepsilon Z_1 D\,, \qquad (3.5.4)$$

$$\varepsilon\dot{D} = \tau^{-1}(-D + \alpha u - \gamma Z_1\dot{\varphi})\,, \qquad (3.5.5)$$

where ε is a small parameter, $\omega^2 = C/J$, $2\varepsilon n = b/J$, $Z_1 = Z/J$, $\gamma = Jk^2\chi(u)/R$, $\alpha = k/R$, $\varepsilon D = Q$.

The transformed functional

$$\Im_d = \int_0^{t_f} D^2(t)dt \qquad (3.5.5)$$

is finite in the final time interval.

Hence we now have a singular perturbed problem for the optimal control voltage. The method of solving such a problem is extensively described in the literature (see, for example, [64, 105]), however the majority of work is related to the study of situations when the sought control depends on direct and conjugate variables. In this case the problem of approximate optimisation leads to adjustments of a form of boundary layer to a solution of boundary problem of the maximum principle.

The use of such a method in the general nonlinear case appears impossible [44, 64, 87] and is effective only when the primary contribution to the functional is introduced by a "slow" (i.e. independent of t/ε) control component. In [44] it is shown that in systems of type (3.5.4) to (3.5.6) the primary contribution to the functional emerged by a "slow" control component and a "fast" component appears to have a significant influence only over the short distance near to the transition point. For this reason approximate solution of (3.5.2) and (3.5.4) to (3.5.6) leads to the selection of "slow" control component.

To explain the peculiarities of resonant motion we will transform equation (3.5.4) into standard form by using the approach described in §3.4. Introducing new variables of "amplitude-phase" according to the expressions

$$\varphi = A\cos\psi,\ \dot{\varphi} = -A\sin\psi, \tag{3.5.7}$$

we obtain

$$\begin{aligned} \dot{A} &= -\varepsilon\omega^{-1}(Z_1 D + 2n\omega A\sin\psi)\sin\psi, \\ \dot{\psi} &= \omega - \varepsilon(\omega A)^{-1}(Z_1 D + 2n\omega A\sin\psi)\cos\psi. \end{aligned} \tag{3.5.8}$$

Considering ψ as a new unknown variable and neglecting values of the second order $\mathcal{O}(\varepsilon^2)$, we transform (3.5.8) into the form

$$\begin{aligned} dA/d\psi &= -\varepsilon\omega^{-2}(Z_1 D + 2n\omega A\sin\psi)\sin\psi, \\ dv/d\psi &= \varepsilon\omega^{-2}A^{-1}(Z_1 D + 2n\omega A\sin\psi)\cos\psi, \end{aligned} \tag{3.5.9}$$

where $v = \omega t - \psi$. The boundary conditions for A, v and the phase ψ_f at the end of the process are defined from the boundary conditions (3.5.2):

$$\begin{aligned} &\psi(0) = 0,\quad \psi_f = \pi - \tan^{-1}(V_f/\omega\varphi_f),\quad v(0) = 0, \\ &v(\psi_f) = \omega t_f - \psi_f,\quad A(0) = \varphi_s,\quad A(\psi_f) = \varphi_f^2 + (V_f/\omega)^2. \end{aligned} \tag{3.5.10}$$

The forced motion of a weakly controlled system is examined in [34] for an interval of time $t_f = \mathcal{O}(\varepsilon^{-1})$. The basic method of study used the averaging method, which allowed the description of the evolution of slow variables for large time intervals. In resonant systems the motion can be assumed to be close to free motion and the cycle of motion is effected in the final interval $t_f = \mathcal{O}(1)$. In this state the parent system is not controlled and has the unique solution $A = \varphi_s$, $v = 0$, while weak control leads to a small, $\mathcal{O}(\varepsilon)$, change in amplitude and phase of the oscillations.

To find the solution we shall use the method of successive approximations [87]:

$$A = \varphi_s + \varepsilon A_1 + \varepsilon^2 \ldots,\quad v = \varepsilon v_1 + \varepsilon^2 \ldots,\quad D = D_0 + \varepsilon \ldots, \tag{3.5.11}$$

where the functions A_1, v_1 and D_0 satisfy the equations

$$dA_1 / d\psi = -\omega^{-2}(Z_1 D_0 + 2n\omega\varphi_s \sin\psi)\sin\psi,$$
$$dv_1 / d\psi = \omega^{-2}\varphi_s^{-1}(Z_1 D_0 + 2n\omega\varphi_s \sin\psi)\cos\psi, \quad (3.5.12)$$

$$\varepsilon dD_0 / d\psi = (\omega\tau)^{-1}(-D_0 + \alpha u - \gamma Z_1 \omega\varphi_s \sin\psi). \quad (3.5.13)$$

Note that in resonant systems the initial and boundary conditions are matched. On account of (3.5.7), (3.5.10) and (3.5.11) we find:

$$\varphi_f = \varphi_s \cos\psi_f + \varepsilon æ, \ A(\psi_f) = \varphi_s - \varepsilon\rho,$$
$$v(\psi_f) = \varepsilon\lambda, \ t_f = \omega^{-1}(\psi_f + \varepsilon\lambda).$$

Below we shall limit the case to that when the arm approaches the latch with low velocity $V_f = \varepsilon^{1/2} V_*$. Then

$$\psi_f = \pi - \varepsilon^{1/2}(V_* / \omega\varphi_f), \ \varphi_f = -\varphi_s + \varepsilon æ. \quad (3.5.14)$$

With the selected accuracy

$$A_1(0) = 0, \ A_1(\psi_f) = -æ + (V_* / \omega^2)/(2\varphi_s) = -\rho,$$
$$v_1(0) = 0, \ v_1(\psi_f) = \lambda. \quad (3.5.15)$$

In [87] it is shown that when using the control voltage u_* in the system (3.5.4) found from the solution of (3.5.12) to (3.5.15) and (3.5.6), the value $\Im_{d*}$ of the functional (3.5.6) will differ from the optimal by a value $\mathcal{O}(\varepsilon^2)$.

Making use of equation (3.5.12) we can write the conditions (3.5.15) in the form of isoperimetric expressions [30, 55]

$$A_1(\psi_f) = -\omega^{-2} \int_0^{\psi_f} [Z_1 D_0(\psi_f) + 2n\omega\varphi_s \sin\psi]\sin\psi d\psi,$$
$$v_1(\psi_f) = \omega^{-2}\varphi_s^{-1} \int_0^{\psi_f} [Z_1 D_0(\psi_f) + 2n\omega\varphi_s \sin\psi]\cos\psi d\psi. \quad (3.5.16)$$

The Hamiltonian function of (3.5.6), (3.5.12) to (3.5.15) takes the form

$$H = -D_0^2 / 2 + (Z_1 D_0 + 2n\omega\varphi_s \sin\psi)(q_1 \sin\psi + q_2 \cos\psi) +$$
$$+ p(\omega\tau)^{-1}(-D_0 + \alpha u - \gamma Z_1 \omega\varphi_s \sin\psi), \quad (3.5.17)$$

where q_1 and q_2 are constant multipliers, p is a Lagrangian multiplier, which satisfy the equations

$$\varepsilon dp / d\psi = -\partial H / \partial D = D_0 - Z_1(q_1 \sin\psi + q_2 \cos\psi) + p(\omega\tau)^{-1},$$
$$p(\psi_f) = 0 \quad (3.5.18)$$

From the maximum principle it follows that

$$u_* = U_0 \eta(p), \quad (3.5.19)$$

where $\eta(p)$ is Heaviside's unit function (2.2.3).

To separate the "slow" component of control, we put $\varepsilon = 0$ in (3.5.13) and (3.5.18). Then

$$\bar{p} = \omega\tau[-\bar{D} + Z_1(q_1 \sin\psi + q_2 \cos\psi)],$$
$$\bar{D} = \alpha U_0 + \bar{\gamma} Z_1 \omega\varphi_s \sin\psi, \quad u_* \neq 0, \tag{3.5.20}$$
$$\bar{D} = 0, \quad u_* = 0.$$

From (3.5.19) and (3.5.20) it follows that

$$\begin{gathered} \text{for the portion of path with } 0 \le \psi_1 < \psi < \psi_2 \le \pi \text{ the control is} \\ u_* = U_0, \quad \bar{D} = \alpha U_0 + \bar{\gamma} Z_1 \omega\varphi_s \sin\psi, \\ \text{for the portions of path with } 0 \le \psi < \psi_1, \text{ and } \psi_2 < \psi \le \pi \\ u_* = 0, \quad \bar{D} = 0. \end{gathered} \tag{3.5.21}$$

The transition points ψ_1, ψ_2 are found from the expressions (3.5.15) and (3.5.16). The curves for u_* and $\bar{D}$ are shown in Fig. 3.5.1. The dashed lines show

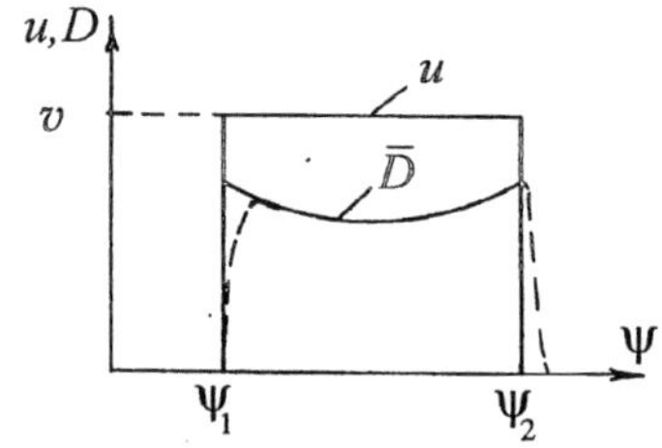

Fig.3.5.1

the smoothing of the initial and final changes in the torque D because of the inertial state of the motor.

In [44] it is proved that when using (3.5.21) the value of the functional with properties (3.5.6) varies from the optimal by no more than $\mathcal{O}(\varepsilon^{1/3})$.

Substituting into (3.5.16) the function $\bar{D}$ from (3.5.21) in place of the function D_0 we obtain, taking account of (3.5.14) and (3.5.15), equations to find the transition points:

$$\begin{aligned} \rho = \omega^{-2}\{\alpha Z_1 U_0(\cos\psi_1 - \cos\psi_2) + \bar{\gamma} Z_1^2 \omega\varphi_s[2(\psi_2 - \psi_1) - \\ -(\sin 2\psi_2 - \sin 2\psi_1)]/4 + \pi n\omega\varphi_s\}, \\ \lambda = (\omega^2\varphi_s)^{-1}[\alpha Z_1 U_0(\sin\psi_2 - \sin\psi_1) - \\ -\bar{\gamma} Z_1^2 \omega\varphi_s(\cos 2\psi_2 - \cos 2\psi_1)/4]. \end{aligned} \tag{3.5.22}$$

The system (3.5.22) requires a numerical solution. For simplicity we shall assume that the duration of the control impulse is small ($\theta = \psi_2 - \psi_1 << 1/2$) and we can take $\sin 2\theta = 2\theta$, $\cos 2\theta = 1$. Then

$$\begin{aligned} \lambda\omega^2\varphi_s = Z_1\theta\cos\psi_l(\alpha U_0 + \bar{\gamma} Z_1\omega\varphi_s \sin\psi_l) = Z_1\theta\bar{D}(\psi_l)\cos\psi_l, \\ r = \omega^2(\rho - \pi n\varphi_s/\omega) = Z_1\theta\bar{D}(\psi_l)\sin\psi_l. \end{aligned} \tag{3.5.23}$$

From (3.5.23) it follows directly that

$$\tan\psi_1 = r(\lambda\omega^2\varphi_s)^{-1}, \quad \theta^2 = [r^2 + (\lambda\omega^2\varphi_s)^2][Z_1\bar{D}(\psi_1)]. \tag{3.5.24}$$

The last equation must satisfy the condition $\theta << 1/2$. Simultaneously the condition $\theta >> \omega\tau$ must be satisfied since in the opposite case it is not possible to neglect effects of the "boundary layer" type [64, 105].

The curve showing the variation of ψ_1 with the system parameters is shown in Fig. 3.5.2. If we consider the parent symmetrical conservative system ($æ = 0,\ V_* = 0,\ n = 0$), then the condition $r > 0$ corresponds to increasing the path

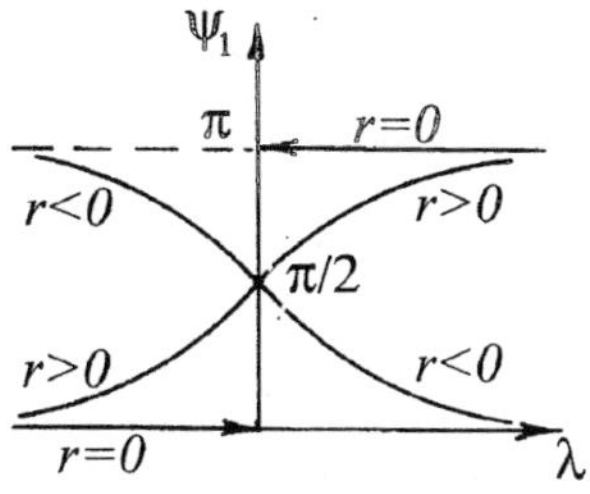

Fig. 3.5.2

length, but the condition $\lambda < 0$ to an accelerating process (by comparison with the parent system). Correspondingly, when $r < 0$ the trajectory is shortened, but when $\lambda > 0$ the process slows down (by comparison with the parent system).

Taking the inequality $\theta > 0$ into account, from (3.5.19) we obtain that $r > 0$ if $\bar{D}(\psi_1) = (\alpha U_0 + \bar{\gamma} Z_1 \omega\varphi_s \sin\psi_1) > 0$ and $r < 0$ if $\bar{D}(\psi_1) < 0$, independently of the position of the transition point ψ_1. The choice of transition point makes it possible to accelerate ($\lambda < 0$) or decelerate ($\lambda > 0$) motion. Specifically, $\lambda \approx 0$ when $\psi_1 \approx \pi/2$ and $r \approx 0$ when $\psi_1 = 0_+$. Thus, as well as in a clock mechanism [2, 25, 61] control near the equilibrium position contributes to a general increase in the sweep of vibration, and does not significantly influence the operating speed. To the contrary, phase shift of the controlling impulse towards the start of the process leads to an increase in the operating speed.

The only variable parameter in the system is the transmission ratio Z_1. Analysing equation (3.5.23) we note that when $\bar{D}(\psi_1) > 0$ there exists a value

$$Z_* = -\alpha U_0 (2\bar{\gamma}\omega\varphi_s \sin\psi_1)^{-1},$$

which minimises the duration of the controlling impulse, but when $\bar{D}(\psi_1) < 0$ the duration of the impulse dies away monotonically as Z_1 increases.

For small values of θ the energy consumption, represented by the functional (3.5.6), takes the form

$$\Im_d \approx \bar{D}^2(\psi_1)\theta = [r^2 + (\lambda\omega^2\varphi_s)^2]^{1/2} \left|\alpha Z_1^{-1} U_0 + \bar{\gamma}\omega\varphi_s \sin\psi_1\right|$$

and decreases continuously as Z_1 increases. It follows that it must be borne in mind that the results obtained are valid when $\theta << 1/2$. Hence when $\bar{D}(\psi_1) > 0$, the term Z_1 must be chosen from the condition $Z_a < Z_* < Z_1 < Z_b$, where Z_a and Z_b are the roots of the quadratic equation $\theta = 1/2$, and when $\bar{D}(\psi_1) < 0$

from the condition $Z > Z_0$, where Z_0 is the positive root of the equation $\theta = 1/2$. If the condition $\theta << 1/2$ is not satisfied then the transition points and duration of the impulse are found by numerically solving the system (3.5.22).

To produce a solution, we have assumed fixed time t_f. As a result we have obtained qualitative and quantitative expressions (3.5.22) for the phases (ψ_1 and ψ_2) and duration of the control impulse for the required operating speed t_f and the actual dissipation n of the drive. However it is obvious that, using a resonant cyclic drive with weak control, can require an operating speed only differing very little from the half-period of vibration of the mechanical system of the drive ($t_f \approx \pi/\omega$). Hence in problems of minimising heat generation in a resonant drive the motion time t_f can be considered without fixation. With optimal control of the system the motion will be little different from free vibration.

In addition, for most resonant drives the velocity of approach of the driven component to the latch may be defined without limits, but the final state of the drive must be set in the form of the functional (3.1.14).

Finally, it is possible not to impose any limit on the required control voltage u, assuming that for implementation of weak optimal control voltage u_* it is not difficult to choose an appropriate motor. The absence of limits on the control voltage makes it possible to obtain optimal control in the form of a simple function when solving an optimisation problem with quadratic functional (3.1.11)

On this basis, we will formulate in a more general form the problem of finding optimal control to ensure minimum heat generation in the resonant drive.

3.5.2. Bolza's problem. The dynamics of a resonant drive with an independently excited direct current electric motor and elastic characteristic $M_1(\varphi)$ for the mechanical energy accumulator can be described by the following equations of motion [22]

$$J\ddot{\varphi} + b\dot{\varphi} + M_1(\varphi) + N_1 \operatorname{sgn}\dot{\varphi} = ZQ,$$
$$Q = [u - Zk\dot{\varphi}]kR^{-1}\chi(u),\ \varphi(0) = \varphi_s,\ \dot{\varphi}(0) = \dot{\varphi}_s, \tag{3.5.25}$$

where variables and parameters are similar to the described before for equations (3.5.1) and (3.5.2).

We shall assume that in the resonant drive the weak control and the dissipative forces are of single order of magnitude $\mathcal{O}(\varepsilon)$. Taking this into account we can rewrite the equations of motion in the form

$$\ddot{\varphi} + M(\varphi) = \varepsilon(D - 2n\dot{\varphi} - N \operatorname{sgn}\dot{\varphi}),$$
$$D = Zks(u - Zk\dot{\varphi}),\ \varphi(0) = \varphi_s,\ \dot{\varphi}(0) = \dot{\varphi}_s, \tag{3.5.26}$$

where $\varepsilon << 1$, $\varepsilon 2n = bJ^{-1}$, $M(\varphi) = M_1(\varphi)J^{-1}$, $\varepsilon N = N_1 J^{-1}$, $\varepsilon D = ZQJ^{-1}$, $\varepsilon s =$ $= (RJ)^{-1}\chi(u)$.

We shall choose φ as a new independent variable. We will assume also that at the start of the cycle, $\dot{\varphi} \geq 0$. Writing $\dot{\varphi} = V$, we can rewrite (3.5.26) in the form

$$dV^2/d\varphi = -2M(\varphi) + 2\varepsilon[Zks(u - ZkV) - 2nV - N],\ V(\varphi_s) = V_s, \tag{3.5.27}$$

The solution of equation (3.5.27) will be sought in the form

$$V^2(\varphi,\varepsilon)=V_0^2(\varphi)+\varepsilon\xi(\varphi,\varepsilon). \qquad (3.5.28)$$

Here $V_0^2(\varphi)$ is the solution of the equation

$$dV_0^2/d\varphi=-2M(\varphi),\ \ V_0(\varphi_s)=V_s, \qquad (3.5.29)$$

which describes the uncontrolled parent motion, $\xi(\varphi,\varepsilon)$ is the solution of the equation of disturbed motion

$$d\xi/d\varphi=2[Zks(u-ZkV)-2nV-N],\ \ \xi(\varphi_s)=0. \qquad (3.5.30)$$

Thus the problem of control of a resonant drive leads to the control of the disturbed component of motion $\xi(\varphi,\varepsilon)$.

The initial problem of finding optimal control can be presented in the form of Bolza's problem [54, 84] by combining the criteria (3.1.1) and (3.1.11) of the functional

$$\Im_e=\min\left\{\mathfrak{G}[\xi(\varphi_f)]+\int_{\varphi_s}^{\varphi_f}D^2V^{-1}d\varphi\right\}, \qquad (3.5.31)$$

with the limits (3.5.28) to (3.5.30), where $\mathfrak{G}[\xi(\varphi_f)]=\beta\,|\,\xi(\varphi_f)-\xi_f\,|$, $D{=}Z\,ks(u-ZkV)$, β is a weighting coefficient, φ_f is the coordinate of the latching point of the driven component. The first term in the functional represents the tolerance on the deviation from the given value of kinetic energy reached for the driven component at the latching position.

Below, we shall limit the search to the first approximation

$$V_1^2(\varphi,\varepsilon)=V_0^2(\varphi)+\varepsilon\xi(\varphi),\ \ \xi(\varphi)=\xi(\varphi,0)$$

and control $u=u_0$, minimising the functional (3.5.31) along the trajectory of the system (3.5.30) when $V=V_0$. The construction of higher approximations requires refinement of the dynamic model of the drive usually taking account of the driver time constant (3.5.1). In the monograph [87] it is shown that when $u=u_0$ the deviation of the system trajectory (3.5.27) and the functional value from the optimal will be of the order of $\mathcal{O}(\varepsilon^2)$.

The Hamiltonian function for problem (3.5.31) when $\varepsilon=0$ takes the form

$$H(\xi,\varphi,p,u)=2p[Zks(u-ZkV_0)-2nV_0-N]-Z^2k^2s^2(u-ZkV_0)^2V_0^{-1}, \qquad (3.5.32)$$

where the Lagrangian multiplier satisfies the equations

$$dp/d\varphi=-\partial H/\partial\xi=0,\ \ p(\varphi_f)=-\partial\mathfrak{G}[\xi(\varphi_f)]/\partial\xi.$$

From this

$$p=\beta\,\mathrm{sgn}\,\sigma, \qquad (3.5.33)$$

where $\sigma=\xi_f-\xi(\varphi_f)$.

The optimal control u_0 is found from the maximum condition

$$u_0=\arg\max_u H(\ldots u)=\arg\max_u\left\{2pu-Zks(u-ZkV_0)^2V_0^{-1}\right\}. \qquad (3.5.34)$$

From this and using (3.5.33) we obtain

$$u_0=(\beta\,\mathrm{sgn}\,\sigma+Z^2k^2s)(Zks)^{-1}V_0(\varphi). \qquad (3.5.35)$$

Then equations (3.5.30) and (3.5.31) can be put in the form

$$d\xi / d\varphi = 2(\beta \operatorname{sgn} \sigma - 2n)V_0 - 2N, \quad \xi(\varphi_s) = 0, \tag{3.5.36}$$

$$\Im_e = \beta\sigma + \beta^2 \int_{\varphi_s}^{\varphi_f} V_0(\varphi) d\varphi. \tag{3.5.37}$$

Thus, to obtain a stable regime of motion for the driven component in a resonant drive with minimum heat generation in the electric motor armature it is essential that the voltage to the driver is maintained proportional to the velocity of the driven component. The coefficient of proportionality depends on dissipation in the drive and the value of the weighting coefficient β. Since weak control causes little change in the mechanical energy of the system, then with error $\mathcal{O}(\varepsilon)$ it is possible to estimate the drive operating speed

$$t_0 = \int_{\varphi_s}^{\varphi_f} V_0(\varphi)^{-1} d\varphi. \tag{3.5.38}$$

Control of a cyclic resonant drive. Using the optimisation approach developed above, we shall derive the control voltage for a resonant manipulator module [4] (Fig. 3.1.1) to ensure the minimum combined functional (3.5.31) for a single displacement with $V > 0$ and boundary conditions $V(-\Delta) = V(+\Delta) = 0$.

To a first approximation, Bolza's problem takes the form

$$\Im_{e1} = \beta_1 \left| \xi(\Delta) - \xi_f \right| + \int_{-\Delta}^{\Delta} Z^2 k^2 s^2 (u - ZkV_0)^2 V_0^{-1} d\varphi, \tag{3.5.39}$$

with the limits

$$dV_0^2 / d\varphi = -2\omega^2 \varphi, \quad V_0(-\Delta) = 0, \tag{3.5.40}$$

$$d\xi / d\varphi = 2[Zks(u - ZkV_0) - 2nV_0 - N], \quad \xi(-\Delta) = 0, \tag{3.5.41}$$

$$V_1^2(\varphi, \varepsilon) = V_0^2(\varphi) + \varepsilon\xi(\varphi), \quad V_1(\Delta) = 0, \tag{3.5.42}$$

where φ is the displacement of the carriage from the stable equilibrium position, $-\Delta$, $+\Delta$ are the coordinates of the latching points, ω is the natural frequency of vibration of the mechanical system.

The parent problem (3.5.40) has the solution

$$V_0^2(\varphi) = \omega^2(\Delta^2 - \varphi^2). \tag{3.5.43}$$

Taking account of (3.5.36) and (3.5.43), equation (3.5.41) for forced motion has the solution [49]

$$\xi(\varphi) = \omega[\beta_1 \operatorname{sgn} \sigma - 2n][\varphi(\Delta^2 - \varphi^2)^{1/2} + \Delta^2(\sin^{-1} \varphi / \Delta + \pi / 2)] - \\ -2N(\Delta + \varphi), \tag{3.5.44}$$

$$\xi(\Delta) = \pi\omega\Delta^2\beta_1 \operatorname{sgn} \sigma - 2\pi\omega\Delta^2 n - 4N\Delta. \tag{3.5.45}$$

The weighting coefficient β_1 can be found from the limit $\xi_f - \xi(\Delta) \le \sigma_1$, imposed on the deviation from the given the value of kinetic energy in the carriage as it approaches the latching position $\varphi_f = +\Delta$. Then replacing the inequality by an equality and substituting (3.5.45) into it, leads to the equation

$$\beta_1 \operatorname{sgn} \sigma_1 = 2n + (4N\Delta + \xi_f - \sigma_1)(\pi\omega\Delta^2)^{-1}. \tag{3.5.46}$$

Substituting (3.5.46) into (3.5.35) and (3.5.37) we obtain the value of the optimal control and the criterion functional

$$u_0 = [2n + Z^2k^2s + (4N\Delta + \xi_f - \sigma_1)(\pi\omega\Delta^2)^{-1}](Zks)^{-1}V_0(\varphi) \tag{3.5.47}$$

$$\Im_{e1} = \beta_1(\sigma_1 + \pi\omega\Delta^2\beta_1 / 2) . \tag{3.5.48}$$

The case $\xi_f = \xi(\Delta) = \sigma_1 = 0$ corresponds to motion of the drive with zero stability margin. As a result control voltage u_0 maintains the mechanical energy in the system (Fig. 3.5.3) unchanged. At the points $\pm\Delta$ the carriage pauses. The

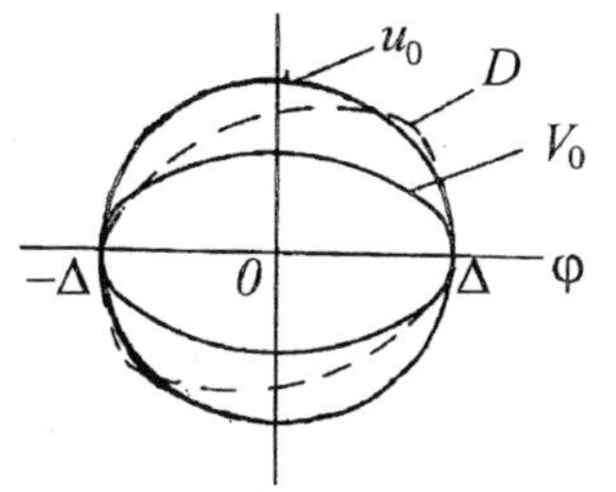

Fig. 3.5.3

operating speed is determined by the natural frequency of the mechanical system $t_0 = \pi\omega^{-1}$. The dotted line in Fig. 3.5.3 shows the variation of torque in the electric motor taking account of its time constant.

Control of a stepped resonant system. We shall derive the optimal control voltage for a weakly controlled multi-position stepped drive module (Fig. 3.1.2), to produce intermediate steps of the driven disk with minimal heat generation in the motor.

As a first approximation, Bolza's problem takes the form

$$\Im_{e2} = \beta_2 \mid \xi(2\pi) - \xi_f \mid + \int_0^{2\pi} Z^2k^2s^2(u - ZkV_0)^2 V_0^{-1} d\varphi , \tag{3.5.49}$$

with limits

$$dV_0^2 / d\varphi = -2M_0(\varphi),\ V_0(0) = V_s , \tag{3.5.50}$$

$$d\xi / d\varphi = 2[Zks(u - ZkV_0) - 2nV_0 - N - M_1(\varphi)],\ \xi(0) = 0 \tag{3.5.51}$$

$$V_1^2(\varphi, \varepsilon) = V_0^2(\varphi) + \varepsilon\xi(\varphi, 0),\ V_1(0) = V_s \geq 0 , \tag{3.5.52}$$

where $M(\varphi) = M_0(\varphi) + \varepsilon M_1(\varphi) = -\omega^2 \sin\varphi - \varepsilon\zeta^2 \sin 2\varphi$, $\omega^2 = 2Cr(a-l)J^{-1}$, $\varepsilon\zeta^2 = 2Cr^2J^{-1}$, $r(a-l)^{-1} << 1$, φ is the deviation of the crank from the unstable equilibrium position, r is crank radius, l and C are initial length and stiffness of the spring respectively, a is the distance between the axes of rotation of the cranks.

The parent problem (3.5.50) has the solution

$$V_0^2(\varphi) = V_s^2 + 2\omega^2(1 - \cos\varphi) . \tag{3.5.53}$$

The first approximations corresponding with (3.5.35), (3.5.51) and (3.5.53) are as follows:

$$\xi(\varphi)=\zeta^2(1-\cos 2\varphi)-2N\varphi+8\omega(\beta_2\,\mathrm{sgn}\,\sigma-2n)[\mathbf{E}(\varphi/2,q)q^{-1}-\\ -\omega\sin\varphi(q'V_0)^{-1}], \tag{3.5.54}$$

$$\xi(2\pi)=-4\pi N+16\omega(\beta_2\,\mathrm{sgn}\,\sigma-2n)\mathbf{E}(q)q^{-1}. \tag{3.5.55}$$

Here $\mathbf{E}(q)$ and $\mathbf{E}(\varphi/2,q)$ are complete and incomplete elliptical integrals of the second order, $q^2=4\omega^2(V_s^2+4\omega^2)^{-1}$, $(q')^2=1-q^2$ are the basic and secondary modules of the Jacobean elliptical functions [1].

The weighting coefficient β_2 is found from the limit $\xi_f-\xi(2\pi)\le\sigma_2$ as for (3.5.46):

$$\beta_2\,\mathrm{sgn}\,\sigma_2=2n+(4\pi N+\xi_f-\sigma_2)q[16\omega\mathbf{E}(q)]^{-1}. \tag{3.5.56}$$

The optimal control voltage and functional are found after substituting (3.5.56) into (3.5.35) and (3.5.37):

$$u_0=\{2n+Z^2k^2s+q(4\pi N+\xi_f-\sigma_2)[16\omega\mathbf{E}(q)]^{-1}\}(Zks)^{-1}V_0(\varphi), \tag{3.5.57}$$

$$\mathfrak{I}_{e2}=\beta_2[\sigma_2+8\omega q^{-1}\mathbf{E}(q)\beta_2]. \tag{3.5.58}$$

The elapsed time of an intermediate step is found in a similar way as in (3.5.38):

$$t_0=2\omega^{-1}q\mathbf{K}(q), \tag{3.5.59}$$

where $\mathbf{K}(q)$ is the complete elliptical integral of first order [1] with modulus q, similar to (3.5.55).

Fig. 3.5.4 shows the phase trajectory and form of the optimal control voltage

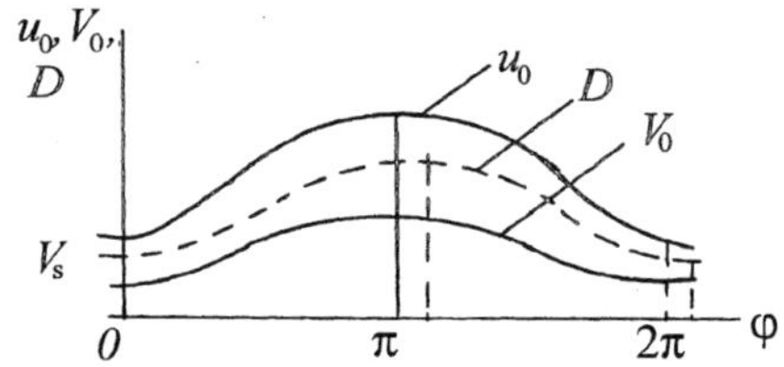

Fig. 3.5.4

for the module over intermediate steps with zero stability margin $\xi_f=\xi(2\pi)=\sigma_2=0$. The control voltage (3.5.57) ensures that the velocity of rotation of the crank is maintained with error $\mathcal{O}(\varepsilon^2)$ at one rotation. Since the number of intermediate steps is known, and V_s is of basic order, the operating regime of the stepped module will be stable.

In Fig. 3.5.4 the dotted line shows the change in torque from the electric motor taking account of its time constant.

When solving the optimal control problem it is quite rare to find analytically the dependence of control voltage from the phase coordinates. However, the knowledge of such dependence facilitates technical implementation of the solutions obtained. The control method of (3.5.35) can be approximately produced by a control system with positive velocity feedback from the driven component (*Patent USSR 1593946, IPC B25J9/00*). The feedback amplification coefficient is chosen on the basis of equation (3.5.36) for excited motion and required stability margin $\xi_f - \sigma$. Obviously the greater the dissipation in the drive and the higher the required stability margin then the larger the feedback amplification coefficient will be.

Fig. 3.1.2 shows an example of technical implementation of quasi-optimal control for a rotating head. The head drive is equipped with the tachogenerator *10*, connected through amplifier *11* to the electric motor *3*. The amplifier arrangement from the initial value calculated from (3.5.57) aims to produce rotation of the disk in intermediate steps with the required stability margin.

§3.6 Control to minimise cumulative angular momentum of the driver

The modular application of resonant drives makes it possible very simply to arrange a robot with multiple degrees of freedom. For example, the MARS resonant robot (Fig. 2.1.1) consists of a rotation module, two linear modules and a stepped module which are successively interconnected. For a machine made in this way, the weight of the modules must be reduced. A significant part of the weight in resonant drives comes from the driving motors. It is well known that the functional criterion (3.1.15), consisting of the cumulative driver angular momentum represents the dimensions and energy consumption of the driver [21]. As distinct from the functionals (3.1.10) to (3.1.12), which have clear energy meaning, the ability to characterise energy consumption by criterion (3.1.15) is not so obvious. Hence, there is interest in comparing the energy efficiency of control obtained using (3.1.15) with that obtained by the functionals (3.1.10) and (3.1.11).

The study of optimal control according to criterion (3.1.15) will be made using the example of the two principal operating regimes for a resonant robot.

The first regime is step-by-step motion (or motion over the first step in multi-stepped motion) in a resonant stepped drive (Fig. 3.1.2) when motion of the driven crank starts with zero velocity from an unstable equilibrium position of the mechanical accumulator $\varphi(0) = 0$, $\dot{\varphi}(0) = 0$, and finishes at a given instant t_f at the same point $\varphi(t_f) = 2\pi$, with velocity limited to $\dot{\varphi}(t_f) = V_f$. To ensure high operating speed requires a more powerful motor generating torque comparable with the inertia forces of the moving components and elastic forces of the spring accumulator. The search for the optimal control voltage can be developed in the form of a solution of Lagrange's problem in which functional (3.1.15) is minimised by limiting the control voltage and fixing t_f and V_f - the time and velocity of approach of the driven component to the latch [21].

The second regime is stable reciprocating/oscillating motion of a cyclic resonant robot drive (Fig. 3.1.1) and stable motion over intermediate steps during multi-step displacement in a resonant stepped drive (Fig. 3.1.2). In these cases weak control is sought by minimising the combination of the functionals (3.1.14) and (3.1.15) (Bolza's problem) over indeterminate time t_f and without any limitation on the control [97].

3.6.1. Lagrange's problem. During motion, the dynamics of a stepped resonant drive with electric motor (1.3.7) can be described by the following differential equations

$$J\ddot{\varphi} + b\dot{\varphi} + M_1(\varphi) = ZQ, \qquad Q = (u - Zk\dot{\varphi})\chi(u)kR^{-1}, \tag{3.6.1}$$

with boundary conditions

$$\varphi(0) = 0, \quad \dot{\varphi}(0) = 0, \qquad \varphi(t_f) = 2\pi, \quad \dot{\varphi}(t_f) = V_f, \tag{3.6.2}$$

Here φ is the angle of deviation of the crank from the unstable equilibrium position; $M_1(\varphi)$ is the elastic characteristic (shown in Fig. 1.4.3, (*a*)) of the mechanical accumulator produced by one of the following structural arrangements: Fig. 1.4.4 (*d*), (*e*); Fig. 1.4.6 (*a*); Fig. 1.4.16, for example using the arrangement Fig. 1.4.6 (*a*) for the stepped module Fig. 3.1.2; t_f is a fixed instant of time. The remaining meanings are similar to those taken in (1.3.7), (3.5.1) and (3.5.2).

We will write the initial systems (3.6.1) and (3.6.2) in the form

$$\dot{\varphi} = V, \ \varphi(0) = 0, \ \dot{\varphi}(0) = 0, \qquad \dot{V} = -M(\varphi) - 2b_0 V + D, \ V(0) = 0, \ V(t_f) = V_f, \tag{3.6.3}$$

where $M(\varphi) = -\omega^2 \sin\varphi - \zeta^2 \sin 2\varphi$, $2b_0 = bJ^{-1}$, $s = (RJ)^{-1}\chi(u)$, $D = ZQJ^{-1} = Zks(u - ZkV)$ and the remaining meanings are similar to (3.5.52).

To reduce the dimensions of the driver and its heat losses in terms of the minimisation functionals chosen the value of the cumulative momentum (3.1.15) is

$$\Im_f = \int_0^{t_f} |D|\, dt. \tag{3.6.4}$$

An additional limit is imposed on the control voltage that $|u| \le U_0$.

To solve problems (3.6.3) and (3.6.4) we shall use the maximum principle. The Hamiltonian function for the system takes the form

$$H(\varphi, V, p_1, p_2, u) = p_1 V + p_2[-M(\varphi) - 2b_0 V + Zks(u - ZkV)] - Zks\,|u - ZkV|, \tag{3.6.5}$$

where the Lagrangian multipliers satisfy the equations

$$\dot{p}_1 = -\partial H / \partial\varphi = p_2 \partial M(\varphi)/\partial\varphi, \qquad \dot{p}_2 = -\partial H / \partial V = -p_1 + p_2(2b_0 + æ_1) - æ_1 \operatorname{sgn}(u - ZkV), \tag{3.6.6}$$

when $æ_1 = (Zk)^2 s$.

The optimal control voltage u_* is found from the maximum condition

$$u_* = \arg\max H(\ldots u) \le U_0 / |u| \le U_0 = \arg\max\{|u - ZkV|[p_2 \operatorname{sgn}(u - ZkV) - 1] / |u| \le U_0\}. \tag{3.6.7}$$

Over the initial stage of motion the driver torque is positive, i.e. in some interval $0 < t < t_1$ we will find $u > ZkV > 0$. Consequently, when

$0 < t < t_{1*},\ t_{1*} \geq t_1$

$$u_* = \arg\max_{|u| \leq U_0} \{(u - ZkV)(p_2 - 1)\} = U_0 \eta(p_2 - 1), \tag{3.6.8}$$

where $\eta(x)$ is Heaviside's unit function (2.2.3). Then, if $U_0 > ZkV$

$$u_* = U_0 \text{ when } p_2 > 1,\ u_* = 0 \text{ when } p_2 < 1. \tag{3.6.9}$$

In what follows, we shall assume that losses in the motor are small and the condition $U_0 > ZkV$ is satisfied over all intervals of motion of the system when $0 \leq \varphi \leq 2\pi$.

Over the braking section when $D \leq 0,\ u \leq ZkV$

$$u_* = \arg\max_{|u| \leq U_0} \{(u - ZkV)(p_2 + 1)\} = -U_0 \eta(-p_2 - 1), \tag{3.6.10}$$

and then

$$u_* = -U_0 \text{ when } p_2 < -1,\ u_* = 0 \text{ when } p_2 > 1. \tag{3.6.11}$$

From (3.6.9) and (3.6.11) it follows that the motor can operate in two modes. If $p_2 > 1$ for all values $0 \leq t \leq t_1$ then on both transition intervals $u = U_0$ and the torque is positive. If $p_2 < -1$ then during the appropriate time intervals a braking torque is generated, $u = -U_0$. Detailed analysis shows that with low energy losses and low velocity V_f, which is representative of actual systems, there must be a braking regime.

From (3.6.9) and (3.6.11) if follows (Fig. 3.6.1) that

Fig.3.6.1

$$u_* = \begin{cases} U_0, & p_2 > 1, \quad 0 < \varphi < \varphi_1, \\ 0, & |p_2| < 1, \quad \varphi_1 < \varphi < \varphi_2, \\ -U_0, & p_2 < -1, \quad \varphi_2 < \varphi < 2\pi. \end{cases} \tag{3.6.12}$$

The transition phase φ_i corresponds to the time instant t_i. The points φ_1, φ_2 (or t_1, t_2) are found from the boundary condition (3.6.2). The interval $0 < \varphi < \varphi_1$ corresponds to acceleration of the system, interval $\varphi_1 < \varphi < \varphi_2$ to free motion produced by the stored energy, and the interval $\varphi_2 < \varphi < 2\pi$ corresponds to a braking torque.

For further analysis it is convenient to choose φ as a new independent variable and using (3.6.12) to rewrite (3.6.3) in the form

$$dV^2/d\varphi = 2[-M(\varphi) - 2b_0V + æu(1-\rho V)],$$
$$V(0) = 0, \quad V(2\pi) = V_f. \tag{3.6.13}$$

where $æ = Zks$, $\rho = æ(sU_0)^{-1}$.

We shall consider the case when the system parameters satisfy the expression $\{b_0, \zeta\} << \omega$ and the condition $U_0 > ZkV$ is satisfied. This occurs when dissipation in the system and losses in the motor are small and the relative deformation of the spring is large by comparison to its length in the undeformed state. With these assumptions equation (3.6.13) can be written in the form

$$dV^2/d\varphi = 2[-M(\varphi) - \varepsilon M_1(\varphi) - 2\varepsilon nV + æu(1-\varepsilon\rho_0V)], \tag{3.6.14}$$

where $M_0 = -\omega^2\sin\varphi$, $M_1(\varphi) = -\zeta^2\sin 2\varphi$, $\varepsilon n = b_0$, $\varepsilon\rho_0 = \rho$.

The features of the system are such that at the instant of latching when $\varphi = 2\pi$ the velocity of the crank is low and we can put $V_f = \varepsilon^{1/2}V_*$. Then the motion (when $\varepsilon = 0$) of the parent system can be described by the equations

$$dV^2/d\varphi = 2[-M_0(\varphi) + æu],$$
$$V_0(0) = V_0(2\pi) = 0. \tag{3.6.15}$$

From (3.6.15) it follows that variation of the velocity by introducing control voltage

$$V_0^2(2\pi) - V_0^2(0) = 2\int_0^{2\pi}[-M_0(\varphi) + æu]d\varphi = 2U_0[\varphi_1 - (2\pi - \varphi_2)] = 0$$

and the transition points are connected by the expressions $\varphi_1 = \theta$, $\varphi_2 = 2\pi - \theta$. Obviously, the higher the velocity $A = V_0(\theta)$ at the end of the acceleration section then the shorter the interval of free motion and, at the same time, to increase A requires a longer acceleration section. We shall show that the total time of motion is decreased when the velocity A increases together with the value of the angle θ.

We find

$$t_f(\theta) = 2\int_0^{\theta} v^{-1}(\varphi)d\varphi + 2\int_{\theta}^{\pi} V_0^{-1}(\varphi,\theta)d\varphi, \tag{3.6.16}$$

where $v(\varphi)$ is the velocity over the acceleration section, $V_0(\varphi,\theta)$ is the velocity over the section of free motion. From (3.6.12) and (3.6.15) we obtain

$$v^2(\varphi) = 2\omega^2(1-\cos\varphi) + 2æU_0\varphi,$$
$$V_0^2(\varphi,\theta) = 2\omega^2(1-\cos\varphi) + 2æU_0\theta. \tag{3.6.17}$$

Introducing (3.6.17) into (3.6.16) and after doing obvious transformation we find

$$dt_f/d\theta = -æU_0\int_{\theta}^{\pi} V_0^{-3}(\varphi,\theta)d\varphi < 0, \quad dt_f/d\theta\big|_{\theta=\pi} = 0. \tag{3.6.18}$$

The inequality (3.6.18) means that the total motion time t_f decreases as the interval when the driver is switched on increases.

Fig. 3.6.2 shows the phase trajectories corresponding to various intervals of switching on the driver ($0<\theta_1<\theta_2<\pi$) for the duration of one rotation $t_{f\infty}>t_{f1}>t_{f2}>t_{f3}$. The minimum time of motion $t_{f3}=t_{\min}$ corresponds to the value $\theta=\pi$ (relay control without switching off the driver). The line when $\theta=0$

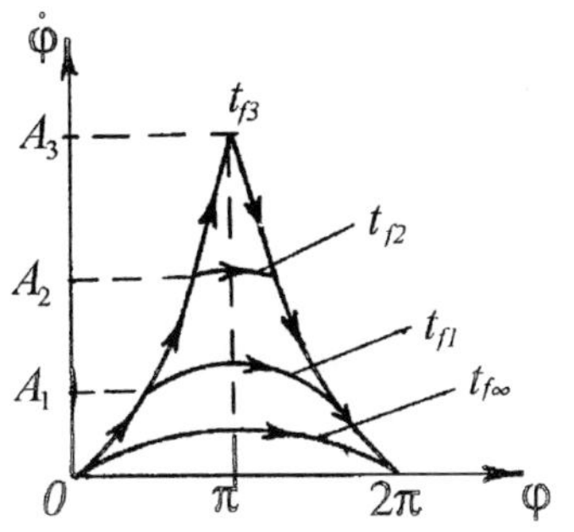

Fig. 3.6.2

(driver not switched in) defines the separatrix with infinite ($t_{f\infty}$) time of motion.

We will assume that the interval when the driver is switched in is small, i.e. the control voltage is used only to generate the identified initial velocity to transfer the system onto the trajectory far from the separatrix. For this we can take $\varphi_1=\varepsilon\theta_1$, $\varphi_2=2\pi-\varepsilon\theta_2$, $U_0=\varepsilon^{-1}U_1$, $\{\theta_1,\theta_2,U_1\}\sim 1$ and rewrite equation (3.6.13) in the form

$$dV^2/d\varphi=2[-M_0(\varphi)-\varepsilon M_1(\varphi)-2\varepsilon nV+\varepsilon^{-1}æu_1(1-\varepsilon\rho_0V)], \qquad (3.6.19)$$

where

$$u_1=U_1,\ 0<\varphi<\varepsilon\theta_1;\quad u_1=-U_1,\ 2\pi-\varepsilon\theta_2<\varphi<2\pi; \qquad (3.6.20)$$
$$V_0(0)=0,\ V_0(2\pi)=\varepsilon^{1/2}V_*.$$

From (3.6.19) and (3.6.20) it follows that during the interval when the motor is switched in $V^2=v_0^2+\mathcal{O}(\varepsilon)$, $\theta_1=\theta_0+\mathcal{O}(\varepsilon)$, where v_0 is the solution of the equation

$$dv_0^2/d\psi=2æU_1,\quad v_0^2=2æU_1\psi, \qquad (3.6.21)$$
$$\psi=\varphi/\varepsilon,\ 0<\psi<\theta_0,\ v_0^2(\theta_0)=A_0^2,\ \theta_0=A_0^2/2æU_1$$

and the influence of the elastic and dissipative elements is significant only in higher approximations. In the same way, the motion of the system during the braking interval $\theta_2=\theta_0+\mathcal{O}(\varepsilon)$ can be described.

In turn, on portions when the driver is switched off, the parent equations take the form

$$dV_0^2/d\varphi=2\omega^2\sin\varphi,\quad \varepsilon\theta_0<\varphi<2\pi-\varepsilon\theta_0, \qquad (3.6.22)$$
$$V_0^2(\varphi,\theta_0)=A_0^2+2\omega^2(1-\cos\varphi)=V_0^2(\varphi).$$

From (3.6.21) the time of motion over the acceleration section ($0<\varphi<\varepsilon\theta_0$) and braking ($2\pi-\varepsilon\theta_0<\varphi<2\pi$) is

$$\tau_p = \int_0^{\varepsilon\theta_0} v_0^{-1}(\varphi)d\varphi = \varepsilon A_0(æU_1)^{-1}. \tag{3.6.23}$$

The time of free motion is

$$\begin{aligned}\tau_0 = \int_{\varepsilon\theta_0}^{2\pi-\varepsilon\theta_0} V_0^{-1}(\varphi)d\varphi = 2\int_0^{\pi} V_0^{-1}(\varphi)d\varphi - 2\int_0^{\varepsilon\theta_0} V_0^{-1}(\varphi)d\varphi = \\ = 2(y_1-\varepsilon y_2),\end{aligned} \tag{3.6.24}$$

$$\begin{aligned} y_1 = 2(A_0^2+4\omega^2)^{-1/2}\mathbf{K}(q_0), \quad q_0^2 = 4\omega^2(A_0^2+4\omega^2)^{-1}, \\ y_2 = \theta_0 A_0^{-1} + \mathcal{O}(\varepsilon) = A_0(2æU_1)^{-1} + \mathcal{O}(\varepsilon).\end{aligned} \tag{3.6.25}$$

Here $\mathbf{K}(q_0)$ is the complete elliptical integral of the first order [1]. Writing the total time of motion over the trajectory in the form $t_f = 2\tau_p + \tau_0$, we obtain

$$t_f = 2y_1 + \mathcal{O}(\varepsilon) = 2\omega^{-1}q_0\mathbf{K}(q_0) + \mathcal{O}(\varepsilon). \tag{3.6.26}$$

In equation (3.6.26) it is not necessary to retain the terms $\mathcal{O}(\varepsilon)$ since these values are neglected when evaluating the velocities v_0, and V_0. Solving (3.6.26) for A_0 with fixed time t_f and substituting A_0 in (3.6.21) we find the phase θ_0 of the motor switching-off.

To take account of losses we will rewrite the equations as a first approximation. From (3.6.19) we have

$$\begin{aligned} dV_1^2/d\varphi = 2[-M_0(\varphi) - \varepsilon M_1(\varphi) - 2\varepsilon n x_0 + \varepsilon^{-1}æu_1(1-\varepsilon\rho_0 v_0)], \\ V_1^2(0) = 0, \quad V_1^2(2\pi) = \varepsilon V_*^2.\end{aligned} \tag{3.6.27}$$

Here $x_0 = v_0$ or $x_0 = V_0$ in the corresponding intervals of motion. Integrating both parts of (3.6.27) over the interval $[0,2\pi]$ and neglecting values of order higher than $\mathcal{O}(\varepsilon)$, we obtain

$$\varepsilon V_*^2 = -4\varepsilon n\int_0^{2\pi} V_0(\varphi)d\varphi - 4\varepsilon æ_1\int_0^{\theta_0} v_0(\varphi)d\varphi + 2æU_1(\theta_1-\theta_2).$$

From this and taking account of (3.6.21) and (3.6.22)

$$\varepsilon\delta = \theta_1 - \theta_2 = \varepsilon(2æU_1)^{-1}[V_*^2 + 32\omega q_0^{-1}\mathbf{E}(q_0) + 8/3æ_1(2æU_1\theta_0^3)^{1/2}], \tag{3.6.28}$$

where $\mathbf{E}(q_0)$ is the complete elliptical integral of the second order [1] with the same parameter q_0 as in (3.6.25).

The equations of the first approximation over the portions of switching in and switching out of the driver take the form

$$dv_1^2/d\psi = 2[æU_1 - 2\varepsilon æ_1 v_0(\psi)], \quad 0<\psi<\theta_1, \quad \psi = \varphi/\varepsilon, \tag{3.6.29}$$

$$\begin{aligned} dV_1^2/d\psi = 2[-M_0(\varphi) - \varepsilon M_1(\varphi) - 2\varepsilon n V_0(\varphi)], \quad \varepsilon\theta_1 < \varphi < 2\pi - \varepsilon\theta_2, \\ V_1(\varepsilon\theta_1) = A_1,\end{aligned} \tag{3.6.30}$$

where v_1 and V_1 are the velocities over the acceleration and free motion sections respectively. The equation of motion over the braking section is similar to (3.6.29). The velocity, playing the role of the primary parameter of the problem, takes the form

$$A_1^2 = v_1^2(\theta_1) = 2æU_1\theta_1 - \varepsilon 4/3æ_1(2æU_1\theta_1^3)^{1/2}. \qquad (3.6.31)$$

It is obvious that the acceleration and braking times can be found from formulas (3.6.23) neglecting the corrections $\mathcal{O}(\varepsilon^2)$. When calculating the time of free motion in (3.6.24) V_0 must be substituted for V_1. In the general case the equation for V_1 is quite complex, however it can be shown that with rapid rotation when $A_0^2 > 4\omega^2(v \sim \varepsilon)$ the solution of equation (3.6.29) can be written as

$$V_1^2 = V_0^2(\varphi, \theta_1) + \mathcal{O}(\varepsilon^2), \qquad (3.6.32)$$

where $V_0(\varphi, \theta_1)$ is found from equation (3.6.22) when A_0 is substituted by A_1. Then by analogy with (3.6.23) to (3.6.26) we can put

$$t_f = 2q_1\omega^{-1}\mathbf{K}(q_1) + \varepsilon A_0(æU_1)^{-1}, \qquad (3.6.33)$$

where $q_1^2 = 4\omega^2(A_1^2 + 4\omega^2)^{-1}$. From (3.6.33) velocity A_1 can be found from which using (3.6.31) and (3.6.28) the transition points are found.

Hence, when the control voltage (3.6.12) $u_*^{(1)} \sim 1$ it is possible to ensure the required operating speed t_f during the first step of motion. The second step begins with the conditions $\varphi = 0$, $\dot{\varphi} = V_f$. As a result the control voltage must only support stable rotation of the crank. The provision of a stable regime of motion can be defined in the form of criterion (3.1.14). In addition to the total torque (3.6.4) the functional (3.1.14) represents a combined criterion of the resonant drive quality (Bolza's problem). Since during motion the driver only compensates for dissipative losses which are not high in a resonant drive, then it is sufficient to have a weak control voltage $u_*^{(i)} \sim \varepsilon$, where $i = 2, 3, \ldots$ - represents the steps in the drive motion. The operating conditions for a cyclic resonant robot are similar.

3.6.2 Bolza's problem. The dynamics of resonant drives with weak control is described by the equations of motion (3.5.25) and (3.5.26). The problem of weak control can be reduced to the problem of controlling the disturbed component of motion $\xi(\varphi, \varepsilon)$ similarly to the way set out in §3.5.2.

The initial problem can be represented in the form of Bolza's problem [54, 84] by a combination of criteria (3.1.14) and (3.6.4) in the functional

$$\Im_g = \min\left\{ \mathfrak{G}[\xi(\varphi_f) + \int_{\varphi_s}^{\varphi_f} |D| V^{-1} d\varphi \right\}, \qquad (3.6.34)$$

where $\mathfrak{G}[\xi(\varphi_f)] = \beta|\xi(\varphi_f) - \xi_f|$, $D = Zks(u - ZkV)$. It is necessary to derive the optimum weak control $u \in [0, U_0]$, which occurs in the system for a unit displacement from the initial condition $\varphi(0) = \varphi_s$, $V(\varphi_s) = V_s$ to the end point φ_f.

The first term in functional (3.6.34) represents the tolerance on the deviation from the given value of the achieved kinetic energy of the driven component in the locked position. The integrated term represents the cumulative angular momentum of the driver.

Later we shall limit the search to the first approximation $V_1^2(\varphi,\varepsilon)=V_0^2(\varphi)+\varepsilon\xi(\varphi,0)$ and control voltage $u=u_0$ to minimise the functional (3.6.34) along the trajectory of the system (3.5.30) when $V=V_0\geq 0$ and $u\in[0,U_0]$.

We shall construct the Hamiltonian for the parent problem

$$H(\xi,\varphi,p,u)=2p[Zks(u-ZkV_0)-2nV_0-N]-Zks\,|\,u-ZkV_0\,|\,V_0^{-1}, \quad (3.6.35)$$

where the Lagrangian multiplier satisfies the equations

$$dp/d\varphi=-\partial H/\partial\xi=0,\ \ p(\varphi_f)=-\partial\mathfrak{G}[\xi(\varphi_f)]/\partial\xi\,.$$

Whence

$$p=\beta\operatorname{sgn}\sigma\,, \quad (3.6.36)$$

where $\sigma=\xi_f-\xi(\varphi_f)$.

The optimal control u_0 is found from the maximum condition

$$\begin{aligned}u_0&=\arg\max_u H(...u)/u\in[0,U_0]=\\&=\arg\max_u\{2pu-|\,u-ZkV_0\,|\,V_0^{-1}/u\in[0,U_0]\}\end{aligned} \quad (3.6.37)$$

In (3.6.36) and (3.6.37) two situations are possible: $p<0$ and $p>0$. If $p<0$ then from the maximum condition it follows that $u_0=0$. Consequently, the condition $p<0$ is satisfied if $\xi_f<\xi_0(\varphi_f)<0$, where $\xi_0(\varphi)$ is the solution of equation (3.6.30) when $u=0$. In other words, if it is allowed that at the instant of locking, a deviation in the kinetic energy is larger than the losses due to dissipation, then the system does not require control. For actual systems, assuming $p>0$ and taking account of the condition $u\geq ZkV$, which means that the drive can not support a generator regime of operation for the electric motor, we obtain

$$u_0=U_0\eta(V_0-V_u)\,, \quad (3.6.38)$$

where $V_u=(2\beta)^{-1}$, $\eta(x)=\begin{cases}0, x\leq 0\\1, x>0\end{cases}$ is the asymmetric unit function [68].

Thus, the optimal control with limited $u\in[0,U_0]$ has a relay form. If $\xi_f>\xi_0(\varphi_f)$ then there will always be an interval when the motor is connected during which $\xi_f\geq\xi(\varphi_f)>\xi_0(\varphi_f)$ and $\sigma>0$, $p>0$. To have a stable operating regime with minimal cumulative angular momentum of the motor then a voltage U_0 must be supplied to the armature windings over the path portion $[\varphi_1,\varphi_2]$, where the velocity of the output component of the energy store is larger than V_u, i.e. in the neighbourhood of a stable equilibrium position. The width of the controlling impulse depends on dissipation in the drive, the value of the weight coefficient β and the required stability margin.

Taking account of (3.6.38), equations (3.6.30) and (3.6.34) can be written in the form

$$d\xi / d\varphi = 2[Zks(u - ZkV_0)\eta(V_0 - V_u) - 2nV_0 - N],\ \xi(\varphi_s) = 0, \tag{3.6.39}$$

$$\Im_g = \sigma(2V_u)^{-1} + ZksU_0 \int_{\varphi_1}^{\varphi_2} V_0^{-1} d\varphi - æ_1(\varphi_2 - \varphi_1), \tag{3.6.40}$$

where $æ_1$ is found from (3.6.6)

Since weak control changes the complete mechanical energy of the system by only a small amount, then with an error $\mathcal{O}(\varepsilon)$ it is possible to estimate the operating speed of the drive from formula (3.5.38).

Using the basic approach of (3.6.35) to (3.6.40) we will derive the optimal weak control voltage for a cyclic resonant module (Fig. 3.1.1) for an isolated displacement with $V > 0$, using the functional (3.6.34).

Bolza's problem, to a first approximation, takes the form

$$\Im_{g1} = \beta_1 \,|\, \xi(\Delta) - \xi_f \,|+ \int_{-\Delta}^{\Delta} Zks(u - ZkV_0)V_0^{-1} d\varphi \tag{3.6.41}$$

with the limits $u \in [0, U_0]$, $u \geq ZkV_0$, $V_0 \geq 0$, and limits (3.5.40) to (3.5.42).

The parent problem (3.5.40) has the solution (3.5.43). Equation (3.5.41) for the disturbed motion taking account of (3.5.43), (3.6.38) has the solution

$$\begin{aligned} \xi(\Delta) = 4ZksU_0\Delta(1-\gamma^2)^{1/2} - 2æ_1\omega\Delta^2\{\gamma(1-\gamma^2)^{1/2} + \cos^{-1}\gamma\} - \\ -2\pi\omega\Delta^2 n - 4N\Delta, \end{aligned} \tag{3.6.42}$$

where $\gamma = V_u(\Delta\omega)^{-1}$.

The weighting coefficient is found from the limit $\xi_f - \xi(\Delta) \leq \sigma_1$, which represents the tolerance on the variation from the required value of the achieved kinetic energy at the latching position $\varphi_f = \Delta$. Then using the equality in the inequality expression and substituting into it (3.6.42) leads to the equation

$$\begin{aligned} 4ZksU_0\Delta(1-\gamma^2)^{1/2} - 2æ_1\omega\Delta^2\{\gamma(1-\gamma^2)^{1/2} + \cos^{-1}\gamma\} = \\ = \xi_f - \sigma_1 + 2\pi\omega\Delta^2 n + 4N\Delta. \end{aligned} \tag{3.6.43}$$

The value of velocity V_u, to achieve which the driver must be supplied with voltage U_0, is found numerically from (3.6.43). The points φ_1, φ_2 at which the driver is switched in or out are found from (3.5.43) when $V(\varphi_1) = V(\varphi_2) = V_u$:

$$\varphi_1 = -\Delta(1-\gamma^2)^{1/2},\ \varphi_2 = -\varphi_1. \tag{3.6.44}$$

The value of functional (3.6.40) assuming optimal control (3.6.38) for the system (3.5.40) to (3.5.42) reaches the value

$$\Im_{g1} = \sigma_1(2V_u)^{-1} + 2ZksU_0\omega^{-1}\cos^{-1}\gamma - 2Z^2k^2s\Delta(1-\gamma^2)^{1/2} \tag{3.6.45}$$

The case when $\xi_f = \xi(\Delta) = \sigma_1 = 0$ corresponds to operation of the module at the threshold of stable motion. When this occurs, the control voltage u_0 maintains the complete mechanical energy of the system (Fig. 3.6.3) unchanged. Since the control impulse is distributed symmetrically about both sides of the equilibrium position, then according to Airey's theorem it has the least effect on the value of the half-period of free vibration of the driven component $t_0 = \pi\omega^{-1}$ [2]. It must be

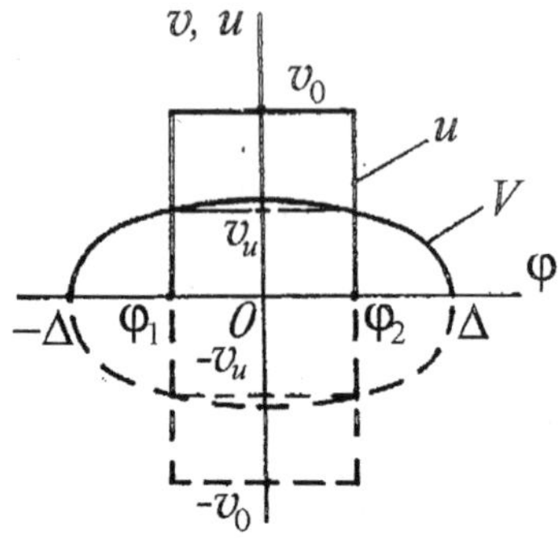

Fig. 3.6.3

noted that for reliable and stable operation of the module it is necessary to have a stability margin. For this reason we choose $\xi_f - \sigma_1 > 0$.

As a result of solving the optimal problems with limits on the control voltage of the form $u \in [0, U_0]$ and functionals (3.5.3) and (3.6.34) the identical control voltages are obtained. This shows that the functional based on the cumulative angular momentum of the motor simultaneously represents the heat losses in the armature windings of the motor and the energy expended in the drive. The conclusive answer to the characteristics of the functional (3.6.34) can be obtained only after an experimental investigation.

We shall construct the optimal weak control voltage for the stepped resonant drive (Fig. 3.1.2), described in §3.1. The control sought must ensure a stable operating regime for the disc *1* over intermediate steps with minimised cumulative angular momentum of the motor.

As a first approximation, Bolza's problem takes the form

$$\Im_{g2} = \beta_2 \,|\, \xi(2\pi) - \xi_f \,| + \int_0^{2\pi} Zks(u - ZkV_0)V_0^{-1} d\varphi\,, \tag{3.6.46}$$

with the limits $u \in [0, U_0]$, $u \geq ZkV_0$, $V_0 \geq 0$, and limits (3.5.50) to (3.5.52).

The parent problem (3.5.50) has solution (3.5.53). A first approximation in agreement with (3.6.38), (3.5.51) and (3.5.53) will be

$$\begin{aligned} \xi(2\pi) = -4\pi N - 32n\omega \mathbf{E}(q)q^{-1} + 4ZksU_0(\pi - \varphi_1) - 16\omega æ_1 q^{-1} \times \\ \times\{\mathbf{E}(q) - \mathbf{E}(\varphi_1/2, q) + \omega q(q'V_u)^{-1}\sin\varphi_1\}. \end{aligned} \tag{3.6.47}$$

Here $\mathbf{E}(q)$ and $\mathbf{E}(\varphi/2, q)$ are complete and incomplete elliptic integrals of the second order, $q^2 = 4\omega^2(V_s^2 + 4\omega^2)^{-1}$, $(q')^2 = 1 - q^2$ are the basic and supplementary modules of Jacobian elliptical functions [1], φ_1 and φ_2 are the points when the driver is switched in or out

$$\varphi_1 = \cos^{-1}[1 - (V_u^2 - V_s^2)(2\omega^2)^{-1}], \quad \varphi_2 = 2\pi - \varphi_1\,. \tag{3.6.48}$$

The weighting coefficient is found from the limit $\xi_f - \xi(2\pi) \leq \sigma_2$ after using the equality

$$\xi_f - \xi(2\pi) = \sigma_2 . \qquad (3.6.49)$$

Then, solving simultaneously equations (3.6.47) to (3.6.49) we can find the coordinate of points φ_1, φ_2 the value of the velocity V_u and the weighting coefficient $\beta_2 = (2V_u)^{-1}$. Due to control voltage (3.6.38) in the system (3.5.50) to (3.5.52) the criterion functional (3.6.46) will take the value

$$\Im_{g2} = \sigma_2 (2V_u)^{-1} + 4ZksU_0 q\omega^{-1}[2\mathbf{K}(q) - \mathbf{F}(\varphi_1, q)] - 2æ_1(\pi - \varphi_1), \qquad (3.6.50)$$

where $\mathbf{K}(q)$ and $\mathbf{F}(\varphi_1, q)$ are the complete and incomplete elliptical integrals of the first order [1].

The time of duration for the intermediate step is found in accordance with (3.5.38): $t_0 = 2\omega^{-1} q\mathbf{K}(q)$.

Fig. 3.6.4 shows the phase trajectory and form of the optimal control voltage

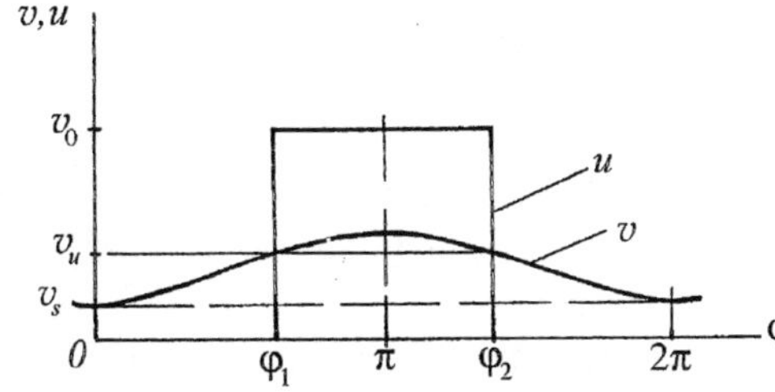

Fig. 3.6.4

for the drive during intermediate steps with zero stability margin during motion $(\xi_f = \xi(2\pi) = \sigma_2 = 0)$. Control voltage u_0 ensures that the speed of rotation of the cranks is maintained over the rotation with error of the order $\mathcal{O}(\varepsilon^2)$. Since the number of intermediate steps is definite, and V_s is of primary order, then the stepped drive will operate in a stable manner.

Control voltage (3.6.38) can be approximated in practice by two methods. The first method is based on establishing switches at the points φ_1 and φ_2. As the drive operates, its moving components interact with the switch that activates or isolates the electric motor. The supply of voltage U_0 to the electric motor can also be provided through relay elements, arranged in the velocity feedback circuit. The threshold of the relay is reached when the velocity reaches V_u, and achieves the required stable operating regime. Obviously, as the dissipative losses in the drive become larger, then the lower the relay threshold and the wider the controlling impulse becomes. Fig. 3.6.5 shows a possible implementation of quasi-optimal control for a stepped module. The drive is provided with a tachogenerator *10* and relay element *11* controlling the switching on and off of the electric motor *3*.

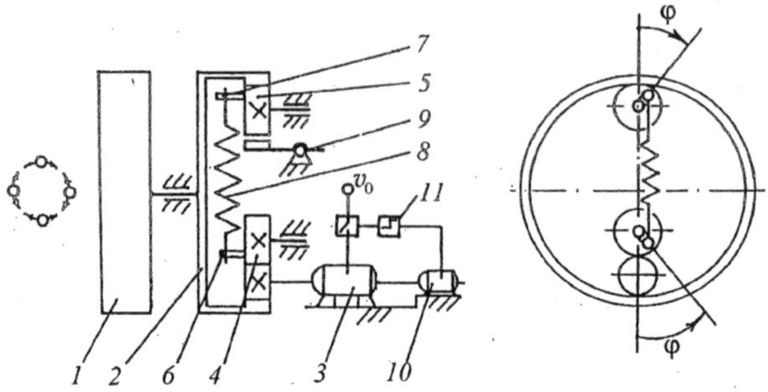

Fig. 3.6.5

§3.7 Efficiency of control for resonant drives

We introduce now a comparative analysis of the efficiency of resonant drive control. To simplify the analytical computation we shall consider control of a cyclic resonant drive (3.3.1) (Fig. 3.1.1) which satisfies the boundary conditions (3.3.2) $\varphi_s = -\varphi_f = -\Delta,\ \dot\varphi_s = \dot\varphi_f = 0$ in the absence of dry friction $N = 0$.

For reference we will take the parameters of the rotation module of the resonant robot MARS [4]: $Z = 12$; $R = 50*$; $J = 0.125 kg.m^2$; $k = 3.2\times10^{-2}$; $\Delta = 1 rad$; $C = 1 Nmrad$; $\omega^2 = 8s^{-2}$.

3.7.1. As mentioned in §3.2 the specific technical complexity requires an energy store with relay elastic characteristic M_1' (3.2.24). With the use of a functional $\Im_a$ (3.2.8) we can evaluate the inefficiency of a drive with energy store consisting of a linear spring having characteristic $M_1'' = \omega^2\varphi$, with a motor controlled according to (3.2.23). Taking into account (3.2.23), the dynamics of the drive is described by the equation

$$\begin{gathered}\ddot\varphi + 2v\tan\ \Delta = 1rad;\ C = 1Nmrad;\tau\dot\varphi + \omega^2\varphi + N = 0,\\ \varphi(0) = -\Delta,\ \dot\varphi(0) = 0,\ \varphi(t_f) = \varphi_f,\ \dot\varphi(t_f) = \dot\varphi_f,\end{gathered} \tag{3.7.1}$$

which, for the case $\dot\varphi_f \ge 0,\ v\omega^{-1} < 1$ has solutions of the form

$$\begin{gathered}\varphi(t) = \frac{N - \Delta\omega^2}{\omega^2\cos v\delta}(v\Omega^{-1}\sin v\tau\sin\Omega\tau + \cos v\tau\cos\Omega\tau) - N\omega^{-2},\\ \dot\varphi(t) = \frac{\Delta\omega^2 - N}{\Omega\cosh v\delta}\cos v\tau\sin\Omega\tau,\end{gathered} \tag{3.7.2}$$

where $\Omega^2 = \omega^2 + v^2$, $v = n[\mu(2n)^{-2} - 1 - \textit{æ}n^{-1}]$, $\tau = \delta + t$, and *æ* is found from (3.2.10). The condition $\dot\varphi(t_f) = \dot\varphi(+\Delta) = 0$ is satisfied by choosing the constant $\delta = -\pi(2\Omega)^{-1}$ and operating speed $t_1'' = \pi\Omega^{-1}$.

The value of the functional criterion (3.2.8) for (3.2.23) and (3.7.2) will be

$$\Im''_{a1} = -4\nu\Delta^2\omega^2 \tan\frac{\pi\nu}{2\Omega}. \tag{3.7.3}$$

When satisfying the conditions $\delta = \varphi_f/2;\ M''_{\max} = \omega^2\varphi_f/2 = M'_0$, where φ_f and M'_0 are from (3.2.33), the ratio of the functional $\Im'_{a1}$ (3.2.34) to $\Im''_{a1}$ (3.7.3) and the operating speeds are practically constant as the value of ν varies within wide limits:

$$\frac{\Im'_{a1}}{\Im''_{a1}} \approx 1.2, \quad \frac{t''_{f1}}{t'_{f1}} \approx 1.1. \tag{3.7.4}$$

Hence an energy store with linear elastic characteristic can be considered as quasi-optimal by comparison to the energy store with relay characteristic (3.2.24).

3.7.2. In §3.3, the optimal control voltage u'_2 (3.3.26) was developed, proportional to the velocity and ensuring minimal total energy consumption by the motor. The complexity of producing an accurate control is due to the fact that it is not easy to construct the feedback linked to velocity with an amplification coefficient dependent on the time of motion.

In §3.5 a quasi-optimal control voltage u''_2 (3.5.47) was derived proportional to the velocity of motion and ensuring minimal heat generation in the motor. Applying criterion (3.3.4) for the total energy consumption we will evaluate the efficiency of the quasi-optimal control u''_2 (3.5.47), based on easily constructed tachometric feedback, by comparison to the accurate optimal control u'_2 (3.3.26). The quasi-optimal control ensures motion of the driven component with zero stability margin $\xi_f - \sigma_1 = 0$ when the motor is supplied with a voltage

$$u''_2 = [2nRJ(Zk)^{-1} + Zk]\dot{\varphi}. \tag{3.7.5}$$

The cycle time with this is determined by the natural frequency of the mechanical

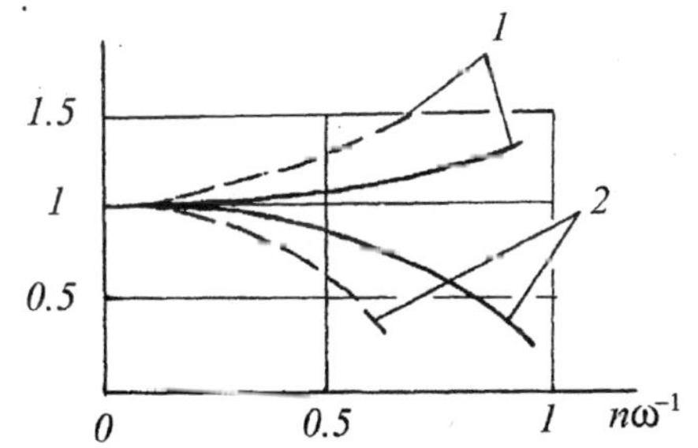

Fig. 3.7.1

system

$$t''_{f2} = \pi\omega^{-1}, \tag{3.7.6}$$

and the functional $\Im_b$ (3.3.4) reaches the value

$$\Im''_{b2} = 2\pi\omega\Delta^2\lambda^2. \tag{3.7.7}$$

where $\lambda = n(1+æn^{-1})^{1/2}$, and $æ$ is defined in (3.2.10). Fig. 3.7.1 shows graphically a comparison between the energy expenditures $\Im'_{b2}$ (3.3.27) and $\Im''_{b2}$ (3.7.7) (curves *1*)

$$\frac{\Im''_{b2}}{\Im'_{b2}} = \pi\lambda\left(2\omega\tanh\frac{\pi\lambda}{2\Omega}\right)^{-1}, \tag{3.7.8}$$

and the periods of motion (curves 2)

$$\frac{t''_{f2}}{t'_{f2}} = (1-\lambda^2)^{1/2}. \tag{3.7.9}$$

The full line shows the curves for the condition $\alpha = 0$ (3.2.10) and the dashed line for $æn^{-1} = 2$. In the figure it can be seen that with damping $n\omega^{-1} < 0.3$ in the mechanical system the results of the effects of control voltages (3.3.26) and (3.7.5) differ little from one another.

3.7.3. With the use of criterion (3.3.4) describing the total energy consumption we will calculate the benefits from using relay energy store M'_l (3.2.24) by comparison to the linear $M''_1 = \omega^2\varphi$ when the driver is controlled by a fixed tachometric feedback according to (3.5.35).

In the system (3.3.1) with $D = 2n\dot{\varphi}$, $N = 0$, $M'_1 = \omega^2\Delta\,\mathrm{sgn}(\varphi-\Delta)$ the total energy expended $\Im_b$ (3.3.4) and the cycle time have the values

$$\Im'_{b3} = \frac{\sqrt{2}}{3}16\lambda^2\Delta^2\omega, \quad t'_{f3} = \sqrt{8}\omega^{-1}. \tag{3.7.10}$$

The same control $D = 2n\dot{\varphi}$ in system (3.3.1) with $N = 0$, $M''_1 = \omega^2\varphi$ ensures the expenditure (3.7.7) and cycle time (3.7.6). As can be seen from the ratios

$$\frac{\Im'_{b3}}{\Im''_{b2}} = 1.2 \quad \text{and} \quad \frac{t'_{f3}}{t''_{f2}} = 0.9, \tag{3.7.11}$$

there is an insignificant gain in operating speed achieved from the use of a relay energy store that worsens the drive economy. Thus it is possible to say that there is an increase in energy expenditure in the control of a resonant drive with relay elastic characteristic (3.2.24).

3.7.4. A well-known method of controlling cyclic resonant drives (*Patent USSR 1346419, IPC B25J9/00)* is one in which the torque supplied to the drive by an electric motor remains constant over the whole range of displacements of the driven component. To achieve this in the drive (3.3.1) with $N = 0$, $M''_1 = \omega^2\varphi$ a control system is used with velocity feedback. The feedback system compensates the emf of the driver $E = Zk\dot{\varphi}$ and the constant reference voltage $u_{ref} = \pi\Delta\omega nRJ(2Zk)^{-1}$ ensures losses due to viscous friction in the system are compensated. Then the general voltage in the driver assumes the value

$$u'_4 = E + u_{ref}. \tag{3.7.12}$$

The cycle time is found from (3.7.6). The energy expenditure of the control calculated in accordance with the functional (3.3.4) for the total energy consumption takes the value

$$\Im'_{b4} = \frac{\pi}{4}\omega\Delta^2 n(n\pi^2 + 8æ) . \qquad (3.7.13)$$

A comparison between the efficiency of controls u''_2 (3.7.5) and u'_4 (3.7.12) shows that at identical cycle time $t'_{f4}/t''_{f2} = 1$, the efficiency of the quasi-optimal control u''_2 (3.7.5) is higher than that of the well-known method

$$\frac{\Im'_{b4}}{\Im''_{b2}} \approx 1.22 . \qquad (3.7.14)$$

3.7.5. To calculate the efficiency of relay controls (3.5.19) from functional (3.5.6) over a fixed time and (3.6.38) from functional (3.6.34) with indeterminate time, we will use the existing dependence of the length of the control impulse $2\varphi_2$ from the limiting control value U_0 supplied to the driver

$$\begin{aligned} u'_5 = U_0 = \pi\omega\Delta^2 nRJ(2Zk\varphi_2)^{-1} + 0.5Zk\omega[(\Delta^2 - \varphi_2^2)^{1/2} + \\ + \Delta^2\varphi_2^{-1}\sin^{-1}(\varphi_2\Delta^{-1})], \end{aligned} \qquad (3.7.15)$$

obtained for the system (3.3.1), which results from equating the friction forces and motive forces over one period of operation, with $N = 0,\ M''_1 = \omega^2\varphi$. The expenditure on control (3.7.15) reckoned from (3.3.4) takes the value

$$\Im'_{b5} = 2\left(\frac{Zk}{RJ}\right)^2 \left\{u_5^2\omega^{-1}\sin^{-1}(\varphi_2\Delta^{-1}) - Zku_5\varphi_2\right\} . \qquad (3.7.16)$$

For $\varphi_2 = \Delta/2$ and with identical cycle times $t'_{f5}/t''_{f2} = 1$ the efficiency of the quasi-optimal control u''_2 (3.7.5) is considerably larger than that for control u'_5 (3.7.15)

$$\frac{\Im'_{b5}}{\Im''_{b2}} \approx 1.67 . \qquad (3.7.17)$$

3.7.6. The simplest and most widely used control for resonant drives consists of supplying a constant control voltage u'_6 to the motor over the entire path of the driven component [4, 6, 7]. To calculate the efficiency we shall use expressions (3.7.15) and (3.7.16) with the condition $\varphi_2 = \Delta$. Then by comparison with the quasi-optimal control voltage u''_2 (3.7.5) we find

$$\frac{\Im'_{b6}}{\Im''_{b2}} \approx 1.35 \text{ and } t'_{f6}/t''_{f2} = 1 , \qquad (3.7.18)$$

where $\Im'_{b6} = \Im'_{b5}(\varphi_2 = \Delta),\ t'_{f6} = \pi\omega^{-1}$.

Note also that

$$\frac{\Im'_{b5}(\varphi_2 = 0.5\Delta)}{\Im'_{b6}} = 1.237 . \qquad (3.7.19)$$

Expression (3.7.19) attests to the low energy efficiency of relay control (3.7.15). Of all the controls considered this appears to be the worst.

3.7.7. The control voltage (3.4.20) obtained from criterion (3.1.6) for the operating speed appears obviously worse from the energy consumption stand point (3.3.4). From (3.2.21) and (3.2.22) it follows directly that even in the complete absence of dissipation $n = N = 0$ in the drive (3.3.1), the optimal control (3.4.20) executes considerable work directed to increasing the operating speed.

An experimental study carried out on the rotation module of a MARS robot with dry friction $N = 0.1Nm$ upheld the analytical conclusions for control efficiency. Table 3.1 presents the results of a comparison of the efficiency of three forms of control: smooth control (3.5.35), control with a constant voltage (3.7.15) when $\varphi_2 = \Delta$, and impulse control (3.7.15) when $\varphi_2 \ll \Delta$; with two functionals: full energy consumption (3.1.10) and heat generation (3.1.11) in the electric motor. From the table it can be seen that more than 80% of the energy consumed by the motor is emitted as heat in its windings. This explains the low efficiency of micro-motors [37].

From the analysis of control efficiency the following conclusions can be drawn.

In order to reduce the effort in producing a cyclic resonant drive possessing high operating speed and low energy consumption, it is necessary to use an energy store with linear elastic characteristic $M_1'' = \omega^2\varphi$, and to control the motor using velocity feedback. The choice of amplification factor in the feedback circuit is made through (3.5.47) and ensures the required stability margin in operation for the drive.

If, when choosing the parameters in the feedback circuit according to (3.5.47), the required voltage u_0 supplied to the motor exceeds the tolerable value U_0 then it follows that the control system must be provided by a constant voltage [4].

Table 3.1

Control	u_{max}, V	$\Im_2 = \int_0^{t_f} uidt$, %	$\Im_3 = \int_0^{t_f} Rk^{-2}Q^2dt$,
	20	100	85
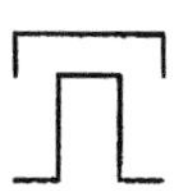	16	113	101
	25	150	138
$100\% = 3.64J$			

Chapter 4. Adaptive control of resonant robots*

§4.1 Introduction

The creation of resonant robots [4, 22, 24] requires the development of motion control concepts for effective maintenance of autoresonant states [16] in spite of changes in parameters of the robot oscillating system. The changing of manipulated mass and the interaction of dynamically coupled degrees of freedom have a major influence on the natural modes of robot oscillation. The dynamical correction of control through the variation of programs or due to the deviation of manipulator parameters during the robot's lifetime is also important.

These parameters and dynamic loads usually cannot be measured preliminarily or directly and have to be identified in an indirect way during the functioning of

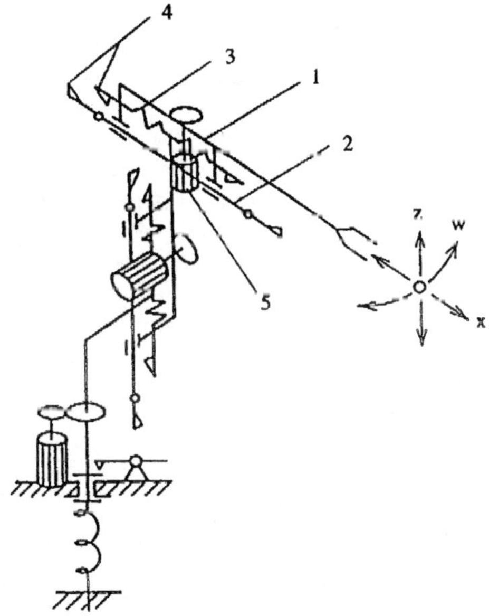

Fig. 4.1.1. Design schematic of resonant robot MARS

the system and evaluated in motion control. Therefore, the times of identification,

* The presentation in this chapter follows that of reference [18].

adaptation and control have to be correlated with the real intervals of cyclic motions. This demands high quality and speed from the proper algorithmic and control structures.

At the same time, the use of resonant control principles permits simplification of the control algorithms, to reduce greatly the energy consumption and to ensure, in this way, an additional resource for increasing control speed and quality.

Fig. 4.1.1 shows the design schematic of the resonant robot MARS [16]. This is a pick-and-place automatic manipulator intended for transportation of light payloads between fixed positions in cylindrical coordinates. The robot consists of the three modules assembled on the "outlet-base" principle in which the moving member of one module serves as attachment for the next one. There is a controlled gripper fixed on the end of the last module.

Each module is an electromechanical auto-oscillating system. It consists of the carriage *1* oscillating along a base guide *2* and connected to the base by springs *3*. The adjustable latches *4* mounted on the base can interrupt the motion at the reversing points. The carriage *1* is connected through gearing to a controllable electric motor *5*.

The module operates in the following way. The carriage is brought into an operating state by fixing one of the latches. To this end, the automatic self-exciting regime is selected by successive reversing of the electric motor according to the direction of movement. When a work stroke is to be made, the latch is automatically withdrawn and the carriage moves to the other extreme position under the force of the spring and motor torque, where it is again fixed by a latch. It retains the energy accumulated by the spring. The next work stroke takes place after reversal of the electric motor and release of the latch.

The programmed cycle of robot operation is accomplished by a controller that establishes the sequence of module moves. The redistribution of energy between spring and carriage takes place during the operation. The total mechanical energy of the module changes itself only because of dissipation and work accomplished.

Since the elastic characteristic of the spring is known, and since the movements carried out are close to the natural modes of oscillation, it is possible to calculate preliminarily the mechanical energy of the module as a whole and to keep this level during motion by means of control, in order to achieve the fixed position with a prescribed kinetic energy (which is necessary for bringing the latch into action).

The problem of resonant module motion control can thus be transformed into the problem of stabilising the prescribed mechanical energy of the system at each point of the trajectory. Depending on the level of parameter uncertainty, and its influence on control quality, it is possible to develop both rough control, which is robust to parameter deviations, and adaptive control. In the latter case, automatic tuning of the control system is fulfilled on-line, as the programmed movements are accomplished. For the simultaneous operation of dynamically coupled modules, it is necessary to compensate additionally for the forces of interaction.

§4.2 Synthesis of regulator

The equation describing the dynamics of unidirectional movement of a resonant robot module, equipped with the spring energy accumulator and a direct current motor with constant moment of inertia and invariable mass of transmission, has the following form:

$$m\ddot{x} + b\dot{x} + F(x - x_0) = u + \alpha mg \qquad (4.2.1)$$

with the boundary conditions

$$x(0) = 0, \quad \dot{x}(0) = 0, \quad x(T) = x_k, \quad \dot{x}(T) = \dot{x}_k,$$

where x is the coordinate of the carriage, x_0 is the coordinate corresponding to the minimum of the spring accumulator energy, $F(x)$ is the elastic characteristic of the spring, $u = nu_d$ is the driving force (u_d is the control voltage, n is the transmission ratio), m is the reduced mass, b is the damping coefficient, α is the coefficient of the gravitational component and dry friction, g is the acceleration due to gravity and T is time of cycle. It is proposed that part of the elastic characteristic disposed symmetrically around the carriage equilibrium position x_0 occurs only during movement. For the calculation of control torque, the static characteristic of the electric motor can be used because its electric time delay is essentially less than T. The permissible velocity $\dot{x}_k^*$ of approaching position x_k is defined as the velocity from the $\varepsilon-$ neighbourhood of nominal value $\dot{x}_k$ chosen from the conditions of reliable fixation.

It is desired that the mechanism with one degree of freedom, equation (4.2.1), produces a control, which transfers the module carriage from the initial position to the target position with a prescribed velocity of final approach and which is invariant to changes in mechanism parameters. The total movement has to be fulfilled during the finite time interval defined by the natural period of carriage oscillation. The control torque is limited.

By proposing that the phase coordinates of the carriage x and $\dot{x}$ are measured permanently, it is possible to make a preliminary calculation in the arbitrary position x of the reference velocity $\dot{x}$ to ensure enough kinetic energy to reach the target position x_k with the prescribed final velocity $\dot{x}_k$

$$\dot{x}_r = \left\{ \dot{x}_k^2 + \frac{2}{\hat{m}}[P(x_k - x_0) - P(x - x_0)] \right\}^{1/2}, \qquad (4.2.2)$$

where $\hat{m}$ is the estimation of moving mass and $P(x)$ is the characteristic of spring energy.

Thus, for the object, (4.2.1), and model, (4.2.2) it is desired to synthesise the control law $|u(x,\dot{x})| \leq U_{max}$ so that, for prescribed t_r and $\varepsilon > 0$, the conditions $|\dot{x}(t) - \dot{x}_r(t)| < \varepsilon$ are ensured for all $t \in (t_r, T)$ and permissible variations of parameters.

The control law for tracing the analytical model (4.2.2) was synthesised as a PI-regulator with additional compensation for dissipation and outer excitation. This type of regulator was chosen due to its robustness. The equation of the regulator is the following:

$$u_d = \hat{b}\dot{x} + k_1 \int_0^t (\dot{x}_r - \dot{x})d\tau + k_2(\dot{x}_r - \dot{x}) + \hat{m}g\hat{\alpha}, \tag{4.2.3}$$

where k_1 and k_2 are tuneable coefficients; $\hat{m}$, $\hat{b}$ and $\hat{\alpha}$ are estimations of parameters received during the identification and tuning.

A control schematic for the module is shown in Fig. 4.2.1. The prescribed velocity $\dot{x}_k$ is introduced as a parameter of the tuneable analytical model (4.2.2). The current coordinate x and velocity $\dot{x}$ are measured during the control time. The co-ordinate x is transmitted simultaneously to the input of the analytical

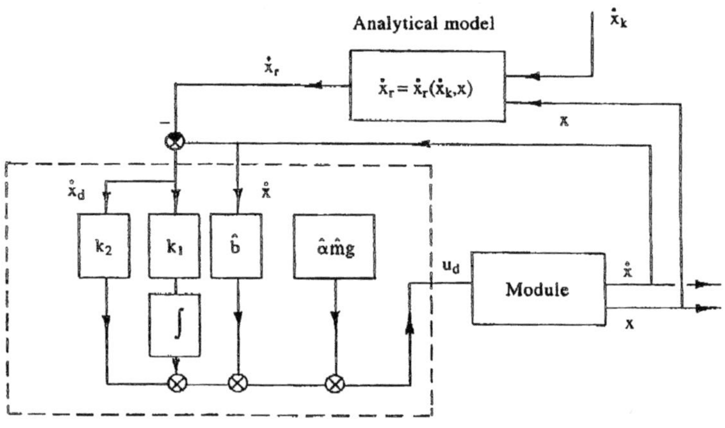

Fig. 4.2.1. Model reference control system.

model. On the basis of received information the model calculates the reference velocity $\dot{x}_r$ corresponding to the current co-ordinate x. The error, $\dot{x}_d = \dot{x} - \dot{x}_r$, is transmitted to the input of the regulator, which also receives the measured signal $\dot{x}$, used to compensate dissipation. The regulator then forms the control voltage u_d for the electric motor according to (4.2.3). The coefficients k_1 and k_2 are tuned to achieve the condition $|\dot{x} - \dot{x}_k| < \varepsilon$ during $t < T$ with an acceptable quality of transient motion and prescribed limitations of control.

The results of the numerical simulation of horizontal translation module dynamics are demonstrated below. The parameters of the module correspond to the real hardware: $m = 0.4kg$, $b = 0.1kg/s$, $\alpha = 0.1$. The elastic characteristic of the spring is linear with stiffness $c = 57N/m$, final velocity $\dot{x}_k = 0$ in the position $x_k = 0.4m$. The control voltage is limited: $|\max u_d| \leq 40V$.

Fig. 4.2.2 shows the curves $\dot{x}_d(t)$ and $u_d(t)$ corresponding to control (4.2.3) with the analytical model (4.2.2) for different values of k_1 and k_2. If

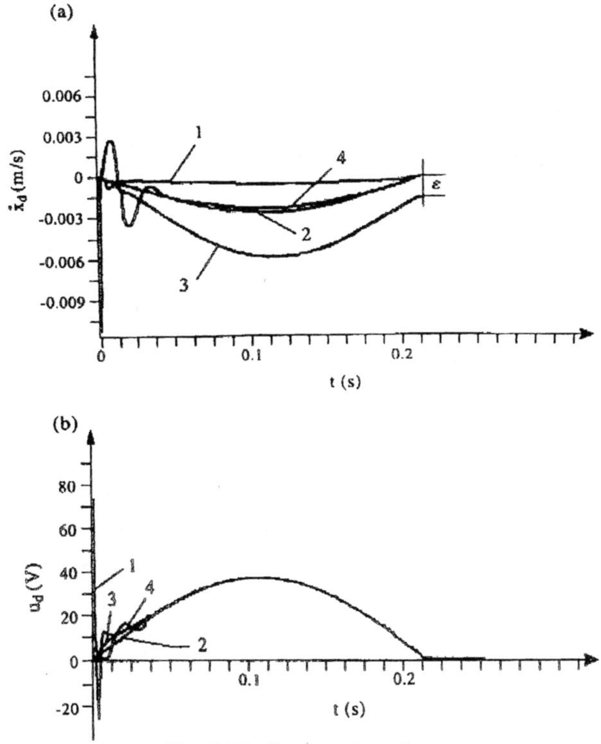

Fig. 4.2.2. Tuning of regulator.
1: $k_1 = 1000,\ k_2 = 100$; 2: $k_1 = 300,\ k_2 = 100$; 3: $k_1 = 20,\ k_2 = 100$; 4: $k_1 = 300,\ k_2 = 20$.
(a) Error of velocity control; (b) control voltage.

$k_1 = 1000,\ k_2 = 100$ (curve 1), the process of regulation is very short, but the control exceeds the prescribed limitation (curve 1, Fig. 4.2.2 *b*). With the decreasing of k_1 to the value $k_1 = 300$, the deflection of the process in the middle of a time interval occurs (curve 2, Fig. 4.2.2 *a*). It is prompted by the high level of perturbation that the integral component has no time to suppress. Meanwhile, by approaching the final position, the velocity $\dot{x}_d$ hits the $\varepsilon-$neighbourhood. The control in this case does not exceed the prescribed limitation (curve 2, Fig. 4.2.2 *b*). Further decrease in k_1 until $k_1 = 20$ leads to the loss of controllability (curve 3, Fig. 4.2.2 *a*).

The tuning of k_2 is performed in a similar way. When $k_2 = 500\ (k_1 = 300)$, the control process is an aperiodic one and the control voltage exceeds the prescribed limitation as in curve 1, Fig. 4.2.2. A very large decrease of k_2 to $k_2 = 20\ (k_1 = 300)$ increases the overshoot (curve 4). As a result of tuning, values of parameters were chosen ($k_1 = 300,\ k_2 = 100$) which permit the fulfilment of a satisfactory control process in the framework of the prescribed limitations.

The robustness of the chosen control strategy was investigated by analysing the control sensitivity to variations in the model parameters. Fig. 4.2.3 demonstrates the results of such analysis. The areas of feasible variations in the parameters are limited by the permissible levels of control voltage. It is seen that the synthesised regulator is robust to deviation of mass, damping and tilting.

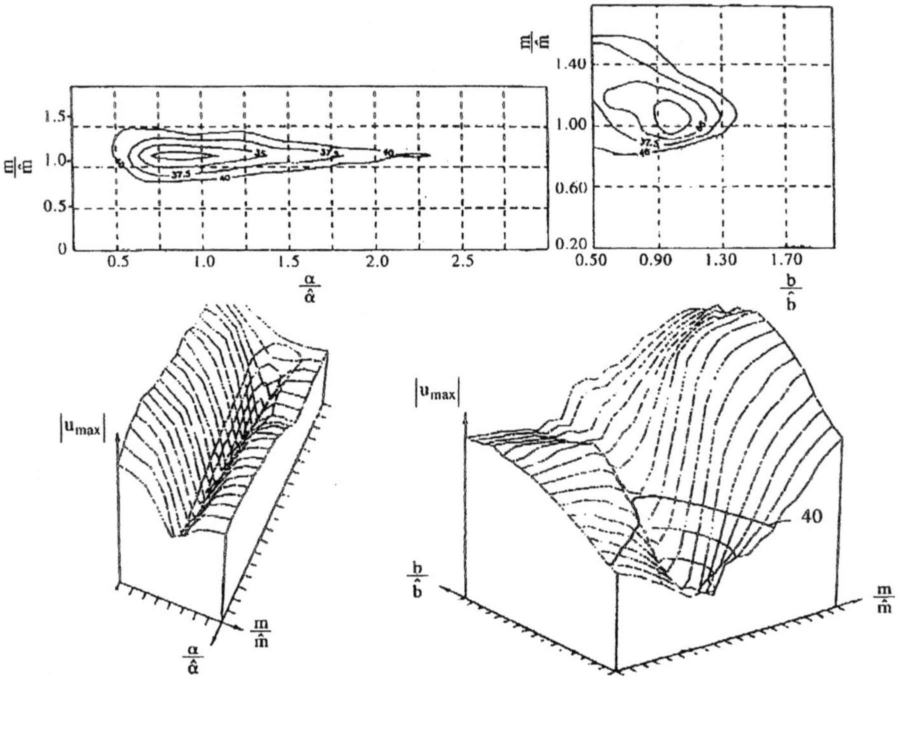

Fig. 4.2.3. Robustness of control (area of feasible parameter variations and control limitations.)

Comparison of the results of simulation with the experimental data received by testing of the horizontal translational module of robot MARS is shown in Fig. 4.2.4. Fig. 4.2.4 *a* corresponds to a comparison of phase trajectories and Fig. 4.2.4 *b* to control voltages. It is seen that the difference between numerical simulation and results of measurement does not exceed 10%. The inaccuracies in the determination of parameter b (curves 2) and parameter α (curves 3) have little influence on the law of control and on the phase trajectories. At the same time, changing the mass by 10% (curve 4) deflects control and phase trajectories substantially.

§4.3 Adaptive control

The regulator described does not ensure that the module functions in the case of substantial deflection of parameters. During operation, the mass of the manipulated payload is usually unknown and can change significantly in every cycle of movement. It emphasises the importance of additional adaptive control that is capable of adjusting the control according to changes in mass.

The adaptive control has to transfer the carriage from the initial to the final position with a prescribed final approach velocity in conditions of unknown moving mass. The total move has to be performed during the finite time defined by the natural period of carriage oscillation, and with limited control voltage.

The general schematic of the development model reference adaptive control

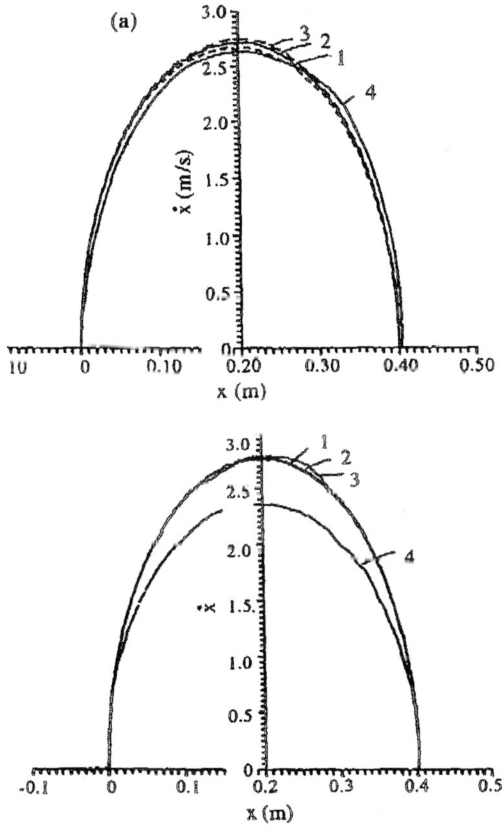

Fig. 4.2.4 (a). Caption overleaf

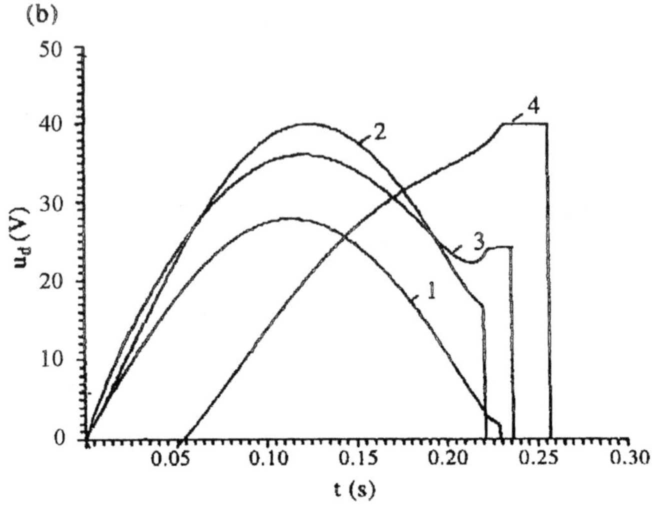

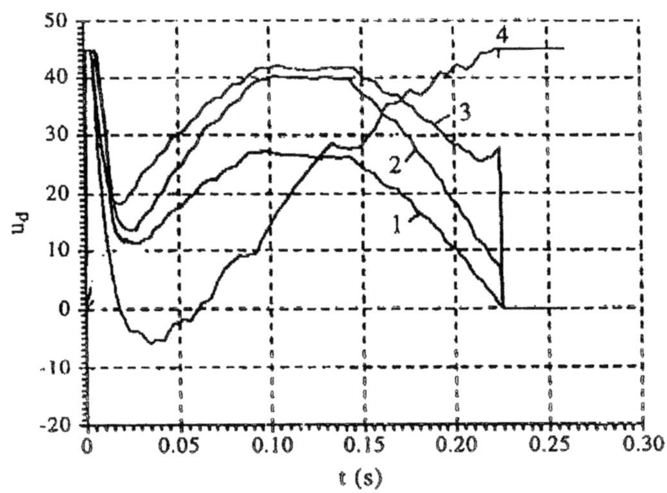

Fig. 4.2.4. Comparison of simulation with experiment.
1: nominal parameters; 2: increment of b ; 3: increment of α ; 4: increment of m .
(a) Phase trajectories; (b) control voltage.

system is shown in Fig. 4.3.1. The process of adaptive control consists of two stages: identification of the unknown parameter m during initial uncontrollable motion and active control of the carriage in accordance with the results of the identification. The use of the separate stages of identification and control simplifies substantially the control process.

During identification, the behaviour of the carriage with unknown mass is compared with the proper output from the model, and the model parameter (mass)

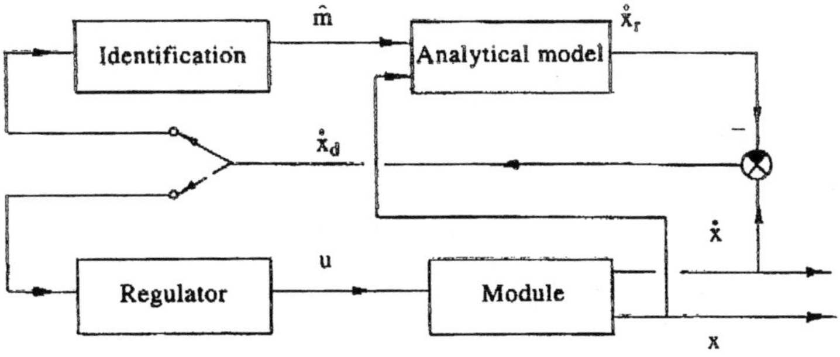

Fig. 4.3.1. Adaptive controls with tuneable model.

is tuned according to comparison error. The value of the parameter that brings the model behaviour into conformity with the real movement of the carriage (with accepted accuracy) is identified as an unknown mass.

Let us define the problem of identification. It is necessary to develop the algorithm of $\hat{m} = m(x, \dot{x})$ estimation in such a manner that, for the prescribed small values Δ and δ, it would exist in such a moment t^* when for all $t > t^*$ the conditions

$$\Im(t) = \int_0^t [\dot{x} - \dot{x}_s(\hat{m})]^2 d\tau < \Delta, \ |\hat{m}(t) - m| < \delta, \ t^* < T$$

would be fulfilled. Here x_s and $\dot{x}_s$ are parameters of the tuneable model described by the equation

$$\frac{\hat{m}\dot{x}_s^2}{2} + P(x_s - x_0) + \hat{b}\int_0^t \dot{x}_s^2 d\tau + \hat{\alpha}\int_0^t g\hat{m}\dot{x}_s d\tau = \frac{\hat{m}\dot{x}_n^2}{2} + P(x_n - x_0), \qquad (4.3.1)$$

where x_n and $\dot{x}_n$ are initial phase co-ordinates. The formulated problem was

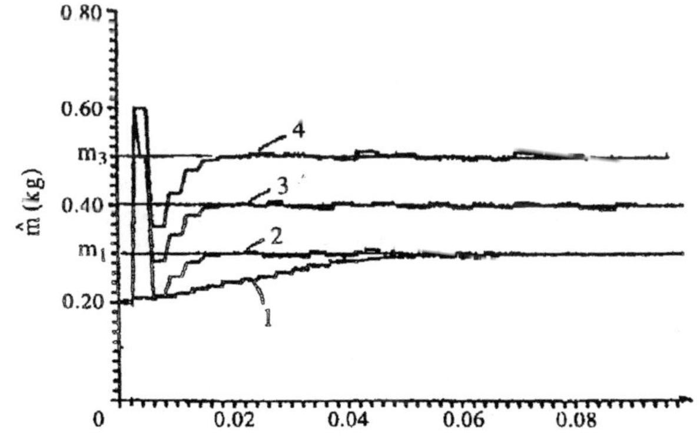

Fig. 4.3.2. Identification of payload mass (influence of γ on convergence)

solved for the described object by the speed gradient method [42]. With this aim the functional $\Im$ is assigned to the function

$$\lambda = \frac{d\Im}{dt} = \lambda(x_s, \dot{x}_s, \hat{m}) . \tag{4.3.2}$$

Then the gradient of this function along the tuneable parameters $\hat{m}$ is calculated and the algorithm of identification is composed according to the rule

$$\dot{\mu} = -\gamma \nabla_\mu \lambda ,$$

where $\mu = \hat{m}^{-1}$, $\lambda = \dot{\Im}$ and γ is the gain coefficient ensuring the desirable speed of convergence.

As a result, the algorithm of carriage mass identification for the resonant module will be the following:

$$\dot{\mu} = -\gamma(\dot{x} - \dot{x}_s) \frac{P(x_s - x_0) - P(x_n - x_0) - \hat{b}\int_0^t \dot{x}_s^2 d\tau}{\dot{x}_s} , \tag{4.3.3}$$

where $\dot{x}_s$ is the tuneable model variable given by the equation

$$\dot{x}_s = \left\langle \dot{x}_0^2 + 2\left\{ \mu \left[P(x_s - x_0) - P(x_n - x_0) - \hat{b}\int_0^t \dot{x}_s^2 d\tau \right] - \hat{\alpha} g(x_s - x_0) \right\} \right\rangle^{1/2} . \tag{4.3.4}$$

It is established that conditions of convergence and identification of this algorithm are satisfied if the estimated parameter is constant and characteristic $P(x)$ is a piece-smooth monotonous function [42]. For the algorithm of (4.3.3) and (4.3.4) presented in the finite-difference form, the condition of choosing coefficient γ ensured that the desirable speed of convergence with prescribed accuracy of estimation was received. The accuracy is limited by the levels of noise in measurement and quantisation.

Fig. 4.3.2 demonstrates the process of mass identification for the different gain coefficients γ and set of payload masses $m_1 < m_2 < m_3$. For the low gain factor γ (curve 1) the process of identification is slow and the sensibility to noise is weak. With an increasing gain factor (curves 2-4) the identification process is accelerated, though there is not total convergence due to the rise of the noise sensibility. It is seen that the duration of identification does not depend on the payload mass.

The functioning of adaptive control was investigated for the parameters of the MARS horizontal translation module. The total algorithm of adaptive control includes initial identification for the tuning of module parameters and the following control with the tuned model. Fig. 4.3.3 demonstrates the results of simulation of the phase trajectories (Fig. 4.3.3 *a*) and control processes (Fig. 4.3.3 *b*) for different durations of identification. It is seen that increasing the identification time in order to obtain a higher estimation accuracy (curve *3*) leads to an increase in object deflection from the nominal trajectory (curve *1*). It causes a later big jump of the control voltage achieving the limitation. The results of simulation (Fig. 4.3.3) were confirmed by experiments (Fig. 4.3.4).

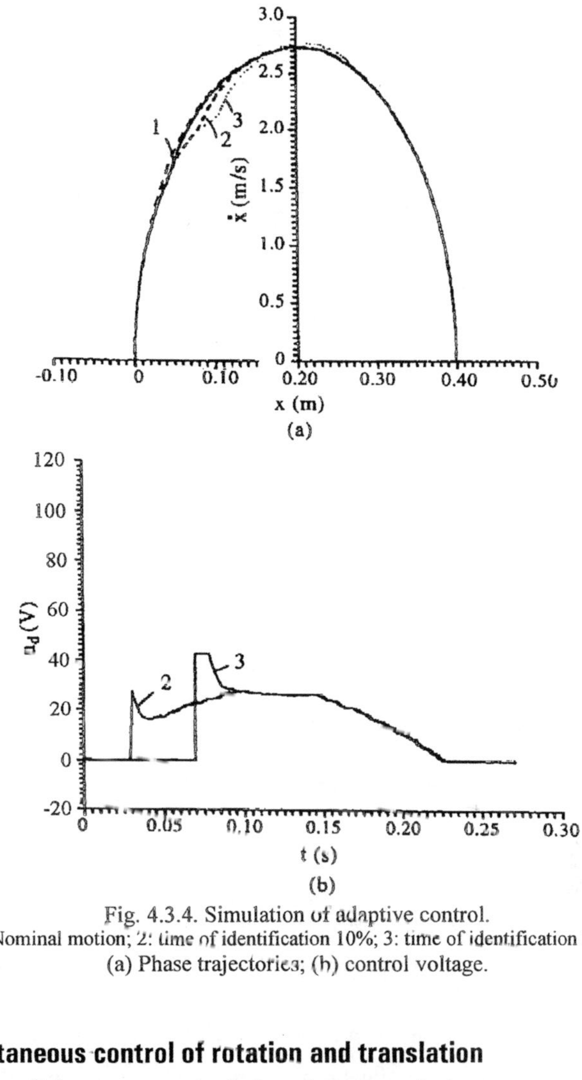

Fig. 4.3.4. Simulation of adaptive control.
1: Nominal motion; 2: time of identification 10%; 3: time of identification 20%.
(a) Phase trajectories; (b) control voltage.

§4.4 Simultaneous control of rotation and translation

To accomplish some manipulation functions, it is necessary to fulfil the simultaneous movement of modules. This leads to the dynamic interaction of rotation and horizontal translation and the control has to overcome this interaction.

The equations in the case of the dynamically coupled movements will be the following (compare with (4.1.1)).

$$m\ddot{x} + b_x\dot{x} + F_x(x - x_0) = m\dot{w}^2 x \operatorname{sgn}\dot{x} + u_x \,, \tag{4.4.1}$$

$$(J_0 + mx^2)\ddot{w} + b_w\dot{w} + F_w(w - w_0) = -2m\dot{x}\dot{w}x + u_w \,. \tag{4.4.2}$$

with the initial conditions

$$x(0) = 0, \quad \dot{x}(0) = 0, \quad w(0) = 0, \quad \dot{w}(0) = 0 .$$

Here x is the co-ordinate of the horizontal module carriage, measured from the axis of rotation, w is the co-ordinate of the turntable, m is the reduced mass of the carriage with payload, J_0 is the initial moment of inertia of the turntable, b_x and b_w are reduced coefficients of viscous friction, $F_x(\ldots)$ and $F_w(\ldots)$ are force characteristics of the module spring elements.

As in the case of independent module movements, let us solve the problem of positioning for every moving element of the module as a problem of tracing its program motion. Each regulator in this case has to bring into line the real and nominal trajectories during the finite time interval under limited control. The final velocity for each module is prescribed by proper values of kinetic energy for reliable response of latches.

In accordance with the method described above for analytical model development, each of the modules has its own analytical model. For the horizontal module it will be

$$\dot{x}_r = \left\{ \dot{x}_k^2 + \frac{2}{\hat{m}} [P_x(x_k - x_0) - P_x(x - x_0)] \right\}^{1/2}, \tag{4.4.3}$$

and for the rotation module

$$\dot{w}_r = \left\{ \frac{2}{J(x)} \left[\frac{J_k \dot{w}_k^2}{2} + P_w(w_k - w_0) - P_w(w - w_0) \right] \right\}^{1/2}, \tag{4.4.4}$$

where $J(x) = J_0 + mx^2$, $J_k = J(x)$.

It is seen from (4.4.4) that, during the calculation of the rotation module velocities, the change of moment of inertia is taken into consideration.

For the synthesis of the control laws in this case, the methods of decomposition, linearisation and compensation of the nonlinearities were used [110]. The equation of the horizontal movement module is

$$u_x = k_1^x \int_0^t (\dot{x}_r - \dot{x}) d\tau + k_2^x (\dot{x}_r - \dot{x}) + \hat{b}_x \dot{x} - \hat{m} \dot{w}^2 x \operatorname{sgn} \dot{x} , \tag{4.4.5}$$

and of the rotational movement module it is

$$u_w = k_1^w \int_0^t (\dot{w}_r - \dot{w}) d\tau + k_2^w (\dot{w}_r - \dot{w}) + \hat{b}_w \dot{w} + \hat{m} \dot{x} x \dot{w}_r \tag{4.4.6}$$

Here k_1^x, k_2^x, k_1^w and k_2^w are regulator coefficients for horizontal and rotation modules; $\hat{m}$, $\hat{b}_x$, and $\hat{b}_w$ are estimations of module mechanism parameters.

It is seen from (4.4.5) and (4.4.6) that in addition to the terms that ensure conformity between model and object output, there are the additional terms to compensate for interaction forces, both centrifugal and Coriolis ones. These forces depend on the motion velocities and substantially exceed the dissipation forces.

Fig. 4.4.1 shows the schematic of the control system developed. Each regulator includes additionally a unit to compensate for interaction forces. The analytical model of rotation now uses the signal of horizontal displacement. The areas of permissible parameter variations were investigated by direct numerical integration, as in the one-dimensional case. The robot parameters were chosen according to the real ones of the robot modification MARS-3 with full payload: $m = 1.9kg$, $b_x = 0.1kg/s$, $c_x = 57\text{N/m}$, $J_0 = 2.4kgm^2$, $b_w = 0.1kgm^2/s$,

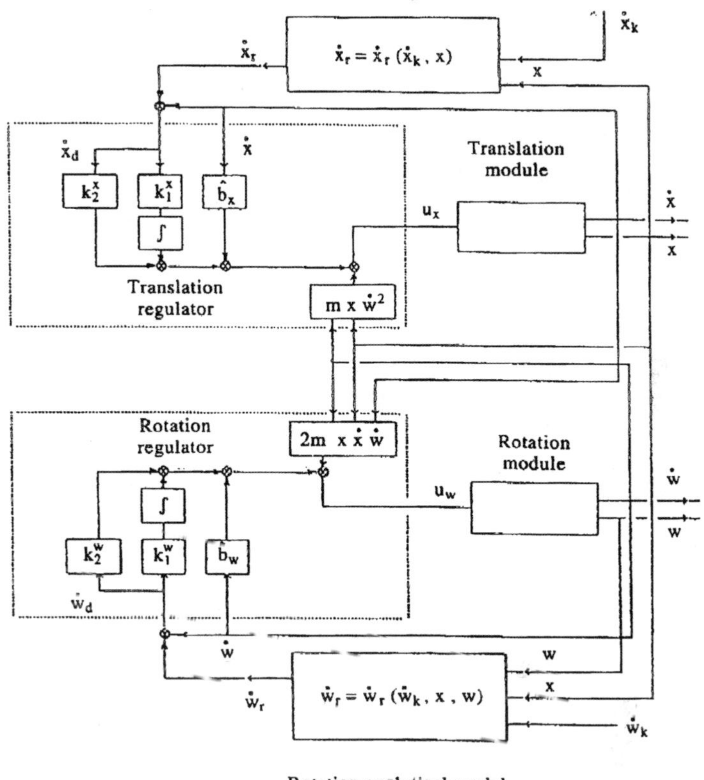

Fig. 4.4.1 Schematic of translation and rotation simultaneous control.

$c_w = 48Nm$. Control voltages were limited in this case as follows: $|\max u_d^x| \leq 100\text{V}$ and $|\max u_d^w| \leq 100\text{V}$.

The regulator coefficients were chosen to ensure control quality: to diminish the transient time and control voltage. As a result, they are as follows: $k_1^x = 350$, $k_2^x = 150$, $k_1^w = 280$ and $k_2^w = 130$. In Fig. 4.4.2 the simulation of simultaneous control is shown. Curve 1 corresponds to translation coordinate x and curve 2 to rotation co-ordinate w.

After the choice of the regulator coefficients the sensitivity of control to variation of model parameters was investigated. As in the case of one-dimensional control, an increased sensitivity to variation of payload mass was observed. Fig. 4.4.3 shows the variations of controls $u_d^x(t)$ and $u_d^w(t)$ depending on the

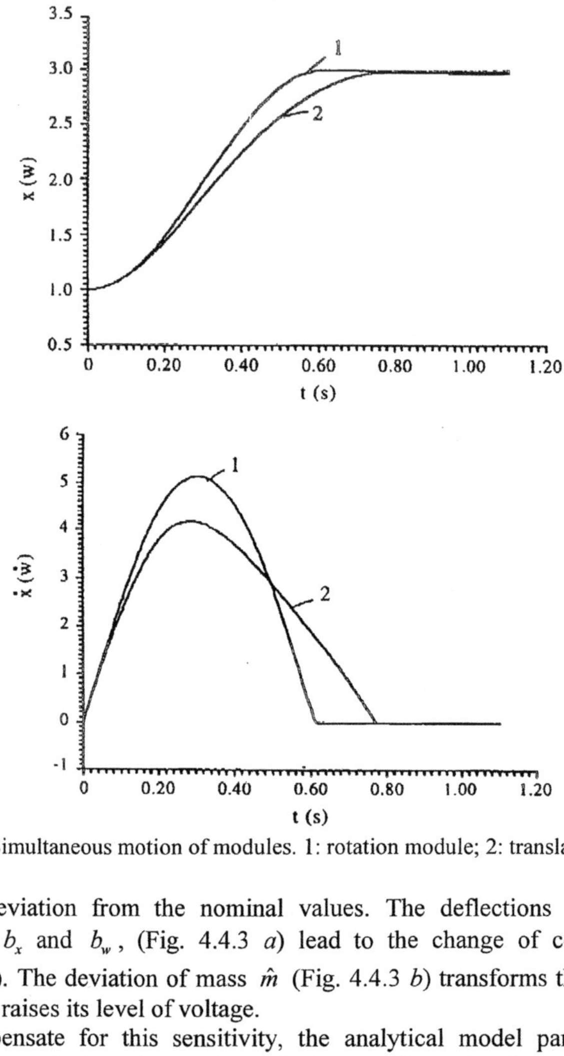

Fig. 4.4.2. Simultaneous motion of modules. 1: rotation module; 2: translation module.

parameter deviation from the nominal values. The deflections of dissipation coefficients b_x and b_w, (Fig. 4.4.3 *a*) lead to the change of control voltage (curves 2, 2'). The deviation of mass $\hat{m}$ (Fig. 4.4.3 *b*) transforms the control and substantially raises its level of voltage.

To compensate for this sensitivity, the analytical model parameters were adapted to the variation of mass $\hat{m}$, as in the one-dimensional control. Fig. 4.4.4 demonstrates the phase trajectories and control voltages in the case of 10% (curves 1) and 15% (curves 2) identification duration. During the initial uncontrolled movements deflection from the nominal partial movements is observed. Switching on the active control causes jumps in the control voltages. Increasing the identification time leads to an increase in these jumps and additional consumption of control energy.

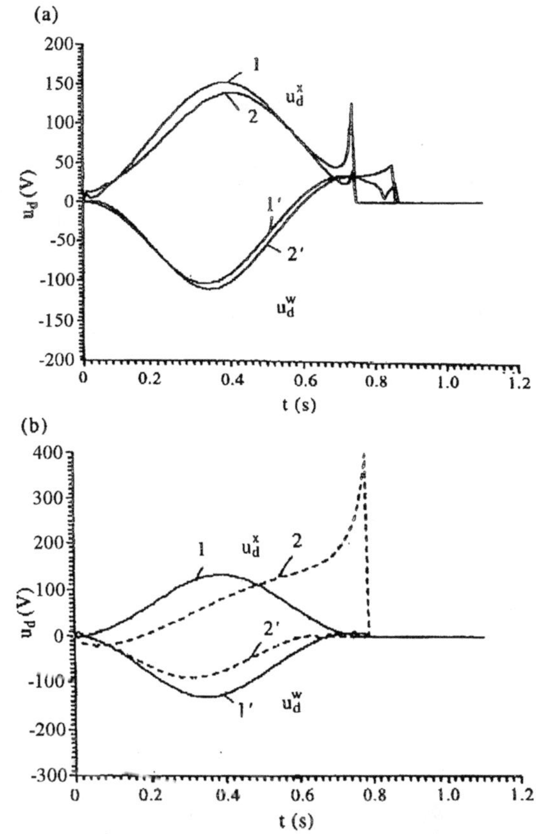

Fig. 4.4.3. Simultaneous control of rotation and translation.
(a) Control of $u_d^x(t)$ and $u_d^w(t)$ at different b_x and b_w.
(b) Control of $u_d^x(t)$ and $u_d^w(t)$ at different m

§4.5 Conclusion

The use of adaptive control principles in the design of resonant robots permits the expansion of their abilities in speed, accuracy and efficiency. The simultaneous motion of dynamically coupled degrees of freedom increases speed of manipulation. The permanent adaptation of control to variations of natural dynamic characteristics of the object leads to the reduction of drive energy consumption. The maintenance of the strict fixation conditions raises the accuracy of positioning and reliability of the robot. It is possible to recognise, through the identification, the type of payload and to change its manipulation program properly.

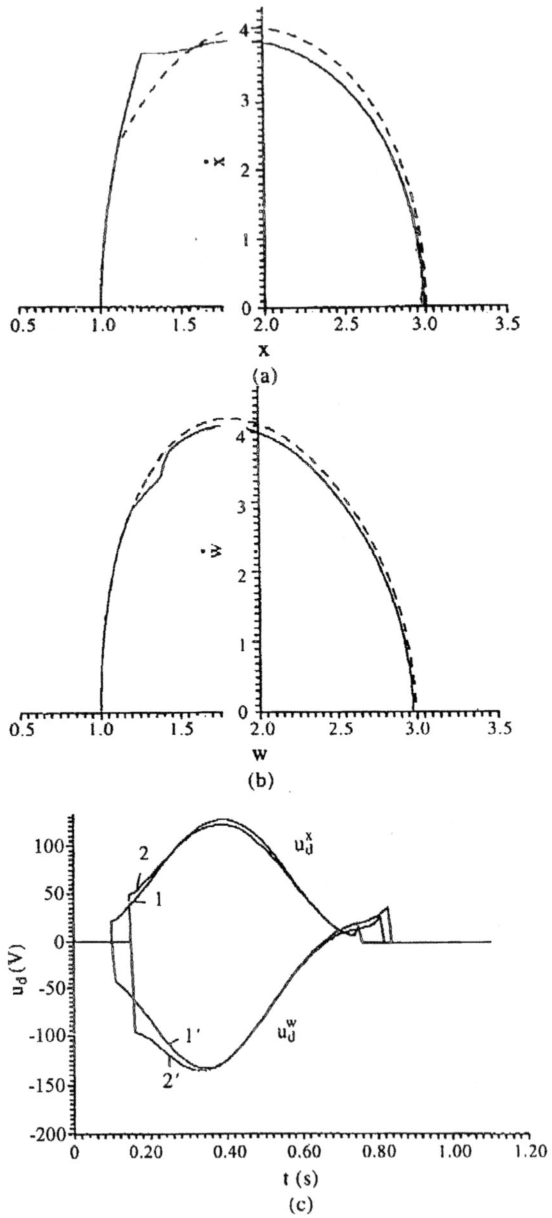

Fig. 4.4.4. Adaptive simultaneous control of rotation and translation.
1: time of identification 10%; 2: time of identification 15%.
(a) Phase trajectory of translation; (b) phase trajectory of rotation; (c) control voltage.

Bibliography

1 Abramowitz M, Stigan I, Handbook of special functions with formulas, diagrams and mathematical tables. Nauka, Moscow, 1979. (In Russian).

2 Airey G, Cambridge Philosophical Transactions, 1827, vol. 2, p. 203.

3 Aizerman M A, Classical mechanics. Nauka, Moscow, 1980. (In Russian).

4 Akynfiev T S, Babitsky V I, Kondratiev V S, Yurchenkov V F, High speed resonant manipulator. Machine Tools and Tools, 1986, no. 2, pp. 9-11. (In Russian).

5 Akynfiev T S, Babitsky V I, Krupenin V L, Resonant-type manipulator systems. Machine Studies, 1982, no. 1 pp. 3-8. (In Russian).

6 Akynfiev T S, Pozharinsky A A, Dynamics of manipulator systems driven by resonant pneumatic drive. Machine Studies, 1984, no. 6, pp. 10-15. (In Russian).

7 Akynfiev T S, Resonant manipulator system with electric drive. Machine Studies, 1983, no. 6, pp. 18-23. (In Russian).

8 Andronov A A, Vitt A A, Khaikyn S E, Theory of oscillations, Dover Publications Inc (Paperback version of translation pub. 1966, Pergamon Press). (Translation from Russian Fizmatgiz, 1959).

9 Artobolevsky I I, Levitsky N I, Cherkudinov S A, Synthesis of plane mechanisms. Gosizdat, Moscow, 1959. (In Russian).

10 Artobolevsky I I, Mechanisms in modern engineering. A handbook in 7 volumes, Nauka, Moscow, 1979-1981. (In Russian).

11 Artobolevsky I I, Theory of mechanisms and machines. Gostekhizdat, Moscow, 1953. (In Russian).

12 Astashev V K, Babitsky V I, Kolovsky M Z, Dynamics and control of machines, Springer, Berlin, 2000.

13 Athans M, Falb P, Optimal Control. McGraw-Hill, New York, 1966

14 Axelrod Z M, Design of clocks and clock systems. Mashinostroenie, Leningrad, 1981. (In Russian).

15 Babitsky V I, Some applications of autoresonant strongly nonlinear vibratory systems. Machine Vibration, 1992, no. 2, pp 110-119.

16 Babitsky V I, Autoresonant mechatronic systems. Mechatronics, 1995, no. 5, pp. 483-495.

17 Babitsky V I, Theory of vibro-impact systems and application, Springer, Berlin, 1998. (Translation from Russian, Nauka, Moscow, 1978).

18 Babitsky V I, Chitayev, Adaptive high-speed robot. Mechatronics, 1996, no. 8, pp. 897-913.

19 Babitsky V I, Kovaleva A S, Control of resonant manipulators to minimise energy consumption. Technical Cybernetics, 1987, no. 3, pp. 108-112. (In

Russian).
20 Babitsky V I, Kovaleva A S, Optimal control of resonant manipulator systems. Machine Studies, 1986, no. 2, pp. 21-25. (In Russian).
21 Babitsky V I, Kovaleva A S, Optimal control of stepped systems of resonant type. Technical Cybernetics, 1988, no. 4, pp. 26-32. (In Russian).
22 Babitsky V I, Kovaleva A S, Shipilov A V, Control of resonant drives with minimal energy consumption. Technical Cybernetics, 1989, no. 4, pp. 210-215. (In Russian).
23 Babitsky V I, Kovaleva A S, Shipilov A V, Optimal control of drives in cyclic machines. Technical Cybernetics, 1989, no. 6, pp. 108-114. (In Russian).
24 Babitsky V I, Kovaleva A S, Shipilov A V, Relieved control in cyclic systems. Arch. Appl. Mech., 1995, no.65, pp. 488-495.
25 Bautin N N, Dynamic theory of clocks. Nauka, Moscow, 1986. (In Russian).
26 Bellman P, Dynamic programming. Princeton University Press, 1957.
27 Besekersky V A, Popov E P, Theory of systems for automatic control. Nauka, Moscow, 1972. (In Russian).
28 Bloch Z Sh, Approximate synthesis of mechanisms. Mashgiz, Moscow, 1948. (In Russian).
29 Bolotnik N N, Chernousko F L, Optimisation of control in robot manipulators. Technical Cybernetics, 1990, no. 1, pp. 189-238. (In Russian).
30 Boltyansky, V G, Mathematical methods for optimal control. Nauka, Moscow, 1966. (In Russian).
31 Bruskin D E, Zorokhovich A E, Khvostov V S, Electrical machines and micro-machines. "Vysshaia Shkola," Moscow, 1981. (In Russian).
32 Chaplais F, Averaging and deterministic optimal control. SIAM J Control and optimiz., 1987, vol. 25, no. 3, pp.767-780.
33 Chen G-J, On the stability of perturbations of optimal control problems. J Math. Anal. Appl 1983, vol. 97, no. 1, pp.46-55.
34 Chernousko F L, Akulenko L D, Sokolov B N, Control of vibrations. Nauka, Moscow, 1980. (In Russian).
35 Chernousko F L, Bolotnik N N, Gradetsky V G, Manipulating robots: dynamics, control, optimisation. Nauka, Moscow, 1989. (In Russian).
36 Chernousko F L, Kolmanovsky V B, Calculation and approximate methods for optimal control. Mathematical analysis, vol. 14. VINITI, Moscow, 1977. (In Russian).
37 Chilikin M G, Klyuchev V I, Sandler A S, Theory of automated electric drives. Energia, Moscow, 1979. (In Russian).
38 Chistov V P, Bondarenko V I, Svyatoslavsky V A, Optimal control of electric drives. Energia, Moscow, 1968. (In Russian).
39 Fedorenko R P, Approximate solutions to the problem of optimal control. Nauka, Moscow, 1978. (In Russian).
40 Filipov I B, Braking arrangements for pneumatic drives. Mashinostroenie, Leningrad, 1987. (In Russian).
41 Fleming W, Rishel R, Optimal control of deterministic and stochastic

systems. Mir, Moscow, 1978. (In Russian).
42 Fomin V, Fradkov A, and Jakubovich V, Adaptive control of dynamic objects. Nauka, Moscow, 1981. (In Russian)
43 Fu K, Gonzalez R, Lie K, Robot engineering. Mir, Moscow, 1989. (In Russian).
44 Gaitzgory V G, On the use of the averaging method to construct sub-optimal solutions of the singular perturbation problem in optimal control. Automation and telemechanics, 1985, no. 9. (In Russian).
45 Gantmakher F R, Lectures on analytical mechanics. Nauka, Moscow, 1966. (In Russian).
46 Genta G, Kinetic energy storage. Theory and practice of advanced flywheel systems. Butterworth-Heinemann, London, 1985
47 Gevondyan T A, Spring drivers. Oborongiz, Moscow, 1956. (In Russian).
48 Gommersall A, Farmer P, Robotics. The international bibliography, with abstracts, Springer, Berlin, 1984.
49 Gradstein I S, Ryzhyk I M, Tables of series, products and integrals. Academic Press, New York, 1965. (Translation from Russian, Fizmatgiz, Moscow, 1963.
50 Gulia N V, Energy storages. Nauka, Moscow, 1980. (In Russian).
51 Gulia N V, Inertial drivers for automobiles. Transport, Moscow, 1974. (In Russian).
52 Gulia N V, Flywheel motors. Mashinostroenie, Moscow, 1976. (In Russian).
53 Guretzky V V, On a problem in optimal control. Proc. AN SSSR Mechanics, 1965, no. 1. (In Russian).
54 Handbook of automatic control, ed. Krasovsky A A. Nauka, Moscow, 1987. (In Russian).
55 Hestens M R, Calculus of variations and optimal control. Wiley Interscience, NY, 1966.
56 Hitrik V F, Methods for dynamical optimisation in automatic machine mechanisms. LGI Leningrad, 1974. (In Russian).
57 Kamke E. Differentialgleichung: Loezungsmethoden und loesungen. Teubner, Stuttgart, 1977
58 Karelin V S, Analytical synthesis of mechanisms - in 2 parts. KPI, Kalinin, 1972-75. (In Russian).
59 Karelin V S, Design of link and toothed link mechanisms. A handbook. Mashinostroenie, Moscow, 1986.
60 Kartvelishvili N A, Galaktionov Yu I, Idealisation of complex dynamic systems. Nauka, Moscow, 1976. (In Russian).
61 Kharkevich A A, Self-sustained vibration. Gostekhizdat, Moscow, 1953. (In Russian).
62 Kobrinsky A A, Kobrinsky A E, Manipulator systems for robots: general arrangements, elementary theory. Nauka, Moscow, 1985. (In Russian).
63 Kobrinsky A E, Dynamics of mechanisms with elastic connections and impact systems. Iliffe Books Ltd., London, 1969. (Translation from Russian, Nauka, Moscow, 1964).
64 Kokotovich P, Application of singular perturbation techniques to control

problems. SIAM Review, 1984, vol. 26, no. 4.
65 Kolovsky M Z, Nonlinear dynamics of active and passive systems of vibration protection. Springer, Berlin, 1999. (Translation from Russian, Nauka, Moscow, 1976).
66 Kolovsky M Z, Dynamics of machines. Mashinostroenie, Leningrad, 1989. (In Russian).
67 Kolovsky M Z, Sloushch A V, General dynamics of industrial robots. Nauka, Moscow, 1988. (In Russian).
68 Korn, G, Korn T, Mathematical handbook for scientists and engineers. McGraw-Hill, New York, 1968.
69 Kovaleva A S, Optimal control of mechanical oscillations. Springer, Berlin, 1999. (Translation from Russian, Nauka, Moscow, 1990.).
70 Kozhevnikov S N, Dynamics of machines with flexible components. Proc. AN SSSR, Kiev, 1961. (In Russian).
71 Kozhevnikov S N, Esipenko Ya I, Raskin Ya M. Mechanisms - a handbook. Mashinostroenie, Moscow. 1976. (In Russian).
72 Kozirev Yu G, Industrial robots - a handbook. Mashinostroenie, Moscow, 1983. (In Russian).
73 Krainev A F, Dictionary/handbook on mechanisms. Mashinostrenie, Moscow, 1981. (In Russian).
74 Kropp A E, Machine drives with impulse variators. Mashinostroenie, Moscow, 1988. (In Russian).
75 Krutko. P D, Control of robot actuators. Nauka, Moscow, 1991. (In Russian)
76 Leech J U, Classical mechanics. Methuen, 1965
77 Levitsky N I, Synthesis of mechanisms following Chebychev. USSR Acad, of Sci., Moscow, 1946. (In Russian).
78 Loitzyansky L G, Lurie A I, A theoretical mechanics course. Vol 1 Statics and kinematics, vol. 2 Dynamics. Nauka, Moscow, 1982-83. (In Russian).
79 Magnus K, Schwingungen. Teubner, Stuttgart, 1976.
80 Maltsev V F, Mechanical impulse transmissions. Mashinostroeneie, Moscow, 1978. (In Russian).
81 Mechanics of industrial robots, a manual for technical universities. Ed. Frolov K V, Vorobev E I. "Vishaia shkola," Moscow, 1988. (In Ru87ssian).
82 Mogilevsky V G, Electromagnetic powder clutches and brakes. Energia, Moscow/Leningrad, 1964. (In Russian).
83 Moiseev N N, Asymptotic methods of nonlinear mechanics. Nauka, Moscow, 1969. (In Russian).
84 Moiseev N N, Calculation methods in the theory of optimal systems. Nauka, Moscow, 1971. (In Russian).
85 Panovko Ya G, Internal friction in elastic system oscillations. Fizmatgiz, Moscow, 1960. (In Russian).
86 Pervozvansky A A, A course in the theory of automatic control: a textbook. Nauka, Moscow, 1986. (In Russian).
87 Pervozvansky A A, Gaitzgory V G. Decomposition, aggregation and approximation of optimisation. Nauka, Moscow, 1979. (In Russian).

88 Petrov B A, Manipulators. Mashinostroenie, Moscow, 1984. (In Russian).

89 Polyudov A N, Programmed releavers for cyclic mechanisms "Vishaia shkola," Lvov, 1979. (In Russian).

90 Pontryagin L S, Boltiansky V G, Gamkrelidze P V, Mischenko E F, Mathematical theory of optimal processes. Wiley Interscience, NY, 1982. (Translation from Russian, Moscow, 1976).

91 Robot control dynamics. Ed. Jurevich E I. Nauka, Moscow, 1984. (In Russian).

92 Sabinin Yu A, Electric machine arrangements for automation. Energoatomizdat, Leningrad, 1988. (In Russian).

93 Schepetilnikov V A, Balancing of mechanisms. Mashinostroenie, Moscow, 1982. (In Russian).

94 Scientific heritage of P L Chebychev: a collection of papers, 2nd edition, ed. Bruevich N G. AN SSSR, Moscow, 1945. (In Russian).

95 Shashkin A S, Toothed lever mechanisms. Mashinostroenie, Moscow, 1971. (In Russian).

96 Shipilov A V, On control of resonant drives with minimal energy expenditure. Problems in machine design and reliability, 1990, no. 3, pp. 93-97. (In Russian).

97 Shipilov A V, Optimal control of resonant drives. Machine Studies, 1989, no. 6, pp. 100-106. (In Russian).

98 Spivakovsky A O, Dyachkov V K, Transport machines. Mashinostroenie, Moscow, 1968. (In Russian).

99 Tikhmenev S S, On one method of producing profiled cams - Detail Design of components and mechanisms for accurate instruments. MVTU, Moscow, 1950. (In Russian).

100 Timoshenko S P, Goodier J N, Theory of elasticity, McGraw-Hill, New York, 1970.

101 Tir K V, Mechanics of automatic printing machines. Kniga, Moscow, 1965. (In Russian).

102 Troitsky V A, Optimal processes in the vibration of mechanical systems. Mashinostroenie, Leningrad, 1976. (In Russian).

103 Troitsky V A, Some problems in the optimisation of manipulator motion. Machine studies, 1988, no. 6, pp. 31-38. (In Russian).

104 Tun A Ya, Methods to control the velocity of electric drives. Energoatomizdat, Moscow, 1984. (In Russian).

105 Vasileva A V, Dmitriev M G, Single perturbations in problems of optimal control, Mathematical analysis, VINITI vol. 20, 1982. (In Russian).

106 Veitz V , Kolovsky M Z, Kochura A E, Dynamics of controlled machine assemblies. Nauka, Moscow, 1984. (In Russian).

107 Vibration in engineering. A handbook in 6 volumes. Mashinostroenie, Moscow, 1981. (In Russian).

108 Volkov A N, Kirillov A V, Lightweight drilling manipulator with elastic accumulator elements and methods for synthesising their parameters, in Mechanising of a working drive with a narrow rock section, Frunze: ILIM, 1978, pp. 54-81.(In Russian).

109 Volkov D P, Krainev A F, Transmissions in construction and road

machines. A handbook. Mashinostroenie, Moscow, 1974. (In Russian).
110 Vukobratovich M, Stokic D, and Kirchansky N, Nonadaptive and adaptive control of robotic manipulators. Mir, Moscow, 1989 (In Russian).
111 Vulfson I I, Dynamic design of cyclic mechanisms. Mashinostroenie, Leningrad, 1976. (In Russian).
112 Wittenburg I, Dynamics of systems of rigid bodies. Teubner, Stuttgart, 1977.
113 Yudovsky I D, Recuperative flywheel drive for fixed function manipulators. Vestnik Mashinostroenia, 1984, no. 5, pp. 9-11. (In Russian).
114 Yukalo P A Analytical synthesis of plane four-bar mechanism. Mashinostroenie, 1981. (In Russian).

Subject Index

Foundations of Engineering Mechanics

Series Editors: Vladimir I. Babitsky, Loughborough University
Jens Wittenburg, Karlsruhe University

Palmov	Vibrations of Elasto-Plastic Bodies (1998, ISBN 3-540-63724-9)
Babitsky	Theory of Vibro-Impact Systems and Applications (1998, ISBN 3-540-63723-0)
Skrzypek/ Ganczarski	Modeling of Material Damage and Failure of Structures Theory and Applications (1999, ISBN 3-540-63725-7)
Kovaleva	Optimal Control of Mechanical Oscillations (1999, ISBN 3-540-65442-9)
Kolovsky	Nonlinear Dynamics of Active and Passive Systems of Vibration Protection (1999, ISBN 3-540-65661-8)
Guz	Fundamentals of the Three-Dimensional Theory of Stability of Deformable Bodies (1999, ISBN 3-540-63721-4)
Alfutov	Stability of Elastic Structures (2000, ISBN 3-540-65700-2)
Morozov/ Petrov	Dynamics of Fracture (2000, ISBN 3-540-64274-9)
Astashev/ Babitsky/ Kolovsky	Dynamics and Control of Machines (2000, ISBN 3-540-63722-2)
Svetlitsky	Statics of Rods (2000, ISBN 3-540-67452-7)
Kolovsky/ Evgrafov/ Slousch/ Semenov	Advanced Theory of Mechanisms and Machines (2000, ISBN 3-540-67168-4)
Landa	Regular and Chaotic Oscillations (2001, ISBN 3-540-41001-5)

Foundations of Engineering Mechanics

Series Editors: Vladimir I. Babitsky, Loughborough University
Jens Wittenburg, Karlsruhe University

Muravskii	Mechanics of Non-Homogeneous and Anisotropic Foundations (2001, ISBN 3-540-41631-5)
Gorshkov/ Tarlakovsky	Transient Aerohydroelasticity of Spherical Bodies (2001, ISBN 3-540-42151-3)
Babitsky/ Krupenin	Vibration of Strongly Nonlinear Discontinuous Systems (2001, ISBN 3-540-41447-9
Manevitch/ Andrianov/ Oshmyan	Mechanics of Periodically Heterogeneous Structures (2002, ISBN 3-540-41630-7)
Lurie	Analytical Mechanics (2002, ISBN 3-540-42982-4)
Slepyan	Models and Phenomena in Fracture Mechanics (2002, ISBN 3-540-43767-3)
Nagaev	Dynamics of Synchronising Systems (2003, ISBN 3-540-44195-6)
Svetlitsky	Statistical Dynamics and Reliability Theory for Mechanical Structures (2003, ISBN 3-540-44297-9)
Neimark	Mathematical Models in Natural Science and Engineering (2003, ISBN 3-540-43680-4)
Babitsky/ Shipolov	Resonant Robotik Systems (2003, ISBN 3-540-00334-7)

Made in the USA
Monee, IL
13 September 2026

62975063R00103